GREAT FEUDS
IN MEDICINE

GREAT FEUDS IN MEDICINE

*Ten of the Liveliest
Disputes Ever*

Hal Hellman

JOHN WILEY & SONS, INC.

New York · Chichester · Weinheim · Brisbane · Singapore · Toronto

Published by John Wiley & Sons, Inc.
Published simultaneously in Canada

No part of this publication may be reproduced, stored in a retrieval system, or transmitted in any form or by any means, electronic, mechanical, photocopying, recording, scanning, or otherwise, except as permitted under Section 107 or 108 of the 1976 United States Copyright Act, without either the prior written permission of the Publisher, or authorization through payment of the appropriate per-copy fee to the Copyright Clearance Center, 222 Rosewood Drive, Danvers, MA 01923, (978) 750-8400, fax (978) 750-4744. Requests to the Publisher for permission should be addressed to the Permissions Department, John Wiley & Sons, Inc., 605 Third Avenue, New York, NY 10158-0012, (212) 850-6011, fax (212) 850-6008, email: PERMREQ@WILEY.COM.

This publication is designed to provide accurate and authoritative information in regard to the subject matter covered. It is sold with the understanding that the publisher is not engaged in rendering professional services. If professional advice or other expert assistance is required, the services of a competent professional person should be sought.

Library of Congress Cataloging-in-Publication Data:
Hellman, Hal
   Great feuds in medicine: ten of the liveliest disputes ever / by Hal Hellman.
      p. ; cm.
   Includes bibliographical references and index.
   ISBN 0-471-34757-4 (cloth : alk. paper)
   1. Medicine—History—Miscellanea.   2. Vendetta—Case studies.   I. Title.
   [DNLM: 1.   History of Medicine, Modern. WZ 55 H478g 2001]
   R133 .H455   2001
   610'.9—dc21

                                                                    00-063349

Printed in the United States of America

10   9   8   7   6   5   4   3   2   1

CONTENTS

Contents

ACKNOWLEDGMENTS

The Internet was helpful. I also made personal visits to several sites: for example, Volta's monument at Lake Como and Golgi's lab at Pavia, both in Italy; the Semmelweis statue in Budapest; and several Freud exhibits in London and New York. But by far the greatest part of my research was done in libraries. Most helpful have been the wonderful collections of historic materials at the New York Academy of Medicine, the New York Public Library, and the newer Science, Industry, and Business Library, all in Manhattan; the Marine Biological Laboratory in Woods Hole, Massachusetts; the American Academy in Rome; and the Burndy Library in Norwalk, Connecticut (now the Dibner Library in Cambridge, Massachusetts).

I would also like to thank the staff at my own local library in Leonia, New Jersey, which is, happily, part of a countywide library system, and through which I was able to reach out and retrieve a remarkable variety of materials from across the country.

A large number of colleagues and friends have been helpful, mainly in answering questions and/or reading and commenting on parts of the manuscript as it was being generated. These include Russell A. Johnson, Archivist and Cataloger, Biomedical Library at the University of California in Los Angeles; Dr. Constance E. Putnam, independent scholar, Concord, Massachusetts; Robert Gallo, M.D., Director, Institute of Human Virology of the University of Maryland in Baltimore; Arthur Peck, M.D., psychiatrist; Dr. Luc Montagnier, Director, Center for Molecular and Cellular Biology, Queens College, City University of New York; Phyllis Dain, Professor Emeritus of Library Science at Columbia University; Morton Klass, Professor Emeritus of Anthropology at Barnard College and Columbia University; Norman Dain, Professor Emeritus of History at Rutgers University; Dr Edward T. Morman, Associate Academy Librarian for Historical Collections, New York Academy of Medicine; Dr. Larry W. Swanson, University of Southern California; Dr. Sonu Shamdasani, Research Fellow,

Wellcome Institute for the History of Medicine; and Leon Hoffman, M.D., New York Psychoanalytic Association.

Additional and special thanks go to my editor, Jeff Golick, who helped me through some difficult times; to my agent, Faith Hamlin, for her support; and especially, to my wife, Sheila, who read every chapter several times, and whose input was invaluable.

INTRODUCTION

In medieval times doctors had only a few ways to determine what was going on inside a patient's body. With no laboratory analysis to aid them, they had to depend on their own senses to obtain a diagnosis. One useful technique was visual examination of the patient's urine. Those who used this method came to be called piss-prophets. Later they would gather additional information by tasting the urine. Sometimes, for obvious reasons, the physician had the patient or even a servant do the tasting.

Uroscopy (diagnostic examination of the urine) eventually went out of fashion. Physicians then did the best they could with other external signs, such as skin and eye color. They also listened to and tried to make sense of the thumping, wheezing, whistling, and crackling sounds sometimes made inside the body, and especially inside the chest and abdomen, which contain the body's major organs. But the activities going on inside that sanctum sanctorum remained mostly a deep, dark secret.

In 1761, Leopold Auenbrugger, a German physician, suggested a way to turn the listening technique from a passive to an active method. His idea was to rap on the patient's thorax, and listen to the echo from the chest cavity. That insistent tap remains a basic tool of the examining physician; who among us has not had his or her chest thumped, and wondered not only what the doctor hears, but what the resulting sounds mean.

Auenbrugger spelled out a variety of different sound responses, and what they might portend. A dull sound correlated with chest congestion, for example, and the duller the sound, the more severe the disease.

Although Auenbrugger believed he had made an important advance, he had few illusions about becoming a medical hero. "In making public my discoveries," he wrote, "I have not been unconscious of the dangers I must encounter, since it has always been the fate of those who have illustrated or improved the arts and sciences by their discoveries to be beset by envy, malice, hatred, destruction and calumny."[1]

We will see examples of each of these in this book. For when any scientist introduces a new theory, he or she is likely to be trampling on someone else's idea. Depending on how well entrenched the original idea is, or how powerful its holder, the responses can take on the virulence expected by Auenbrugger.

In some cases, the attacks seriously disrupted the discoverers' lives. Claude Bernard (chapter 4) worked under constant accusations by antivivisectionists that his physiological experiments on animals were a crime against nature and society. He was even disowned by his own family.

Ignaz Semmelweis, who argued that obstetricians should wash their hands before delivering babies, was dismissed from his hospital job, and ended up in a mental hospital (chapter 3).

Some objections, to be fair, were not entirely unreasonable, and our feuds will show some interesting twists. Semmelweis did indeed end up in a mental hospital, but it's not entirely clear that the actual cause was the treatment meted out by his peers. At least one historian of science argues that his own actions were partly to blame for his sad end.

Another example of not entirely unreasonable objections is seen in the case of René Laënnec. It was Laënnec who had the idea of the stethoscope, which has become another major tool in diagnosis. His idea also faced strong objections, not all of which were foolish; another physician argued: "You will learn nothing by it and, if you do, you cannot treat disease the better." Medical historian Brian Inglis says that this critic "was wrong about there being nothing to learn from auscultation [listening for sounds in the body]: for purposes of diagnosis and prognosis it was invaluable. But he was largely right that it could do little to improve the treatment of disease, which had to await better knowledge of disease processes." [2]

And herein lies the difference between medical practice and medical science; in this book we distinguish carefully between the two. Medical practice is at least as much art as it is science, for physicians are dealing with the most complex structure on the earth. And they are attempting to put to work knowledge and procedures that have been developed by others—researchers who deal in biology, chemistry, engineering, mathematics, statistics, and a variety of other disciplines. Medical science, on the other hand, has laid the basic groundwork that has made possible whatever successes modern medicine has seen. What a glorious, astonishing search it has been.

As in any science, the quest for knowledge has been a driving factor. When, however, the science has, or may have, connection with human health—as is the case in medical science—then the researcher who comes up with a new idea may feel an added urgency.

But, of course, so does the holder of the entrenched method, which the new one may be trying to overthrow. And so it is no surprise that there has been plenty of controversy in the history of medical science.

The reactions feared by Auenbrugger—envy, fear, and so on—are bad enough. But what he got was even worse, and points up the difference between resistance and attack. His technique was not denounced; it was simply ignored, until championed by Corvisart, physician to Napoléon, several decades later.

Auenbrugger would probably have been better off if he had been denounced, or if his method had been attacked. As most advertising copywriters will agree: "I don't care what they say about us—as long as they spell our name correctly."

In fact, one of the points I hope to make in this book is that scientific controversy can have beneficial effects as well as negative ones. The controversy may very well bring the subject into a more public arena, promoting discussion. Getting to the truth may still take time, and the participants may take a beating, but the outcome is likely to be quicker than with a development that leads mainly to silence.

Ongoing controversy can have another beneficial effect: the participants themselves may prosper. Pasteur, for example, thrived in the midst of one battle after another (chapter 5). Not only did he love the fight but, in some cases, he was forced to probe even deeper into the problem at hand and thereby came up with greater advances.

It is worth delving into these controversies for another reason: Often what is behind an attack is some sort of subtle, or not so subtle, driving force. Examples include religion (Harvey, chapter 1) and nationalism (Golgi, an Italian, versus Ramón y Cajal, a Spaniard, chapter 6; and Gallo, an American, versus Montagnier, a Frenchman, chapter 10).

Harvey's story also shows how intertwined were the science, religion, and mysticism of his time. Viscount Conway advised his daughter-in-law not to use Harvey as her physician; he felt that "to have a Physitian abound in phantasie is a very perilous thing. . . ."[3]

Another fertile area for controversy has to do with priority disputes. While simultaneous discovery seems surprising, it is actually fairly common. Well-known examples include Faraday and Henry (electromagnetic induction), Newton and Leibniz (discovery of the calculus), Adams and Leverrier (discovery of Neptune), Darwin and Wallace (theory of evolution), Heisenberg and Schrödinger (quantum mechanics), Schally and Guillemin (thyroid-stimulating hormone), and, in this book, Gallo and Montagnier.

It is true that one of the major drives in any science is just the pleasure of finding things out, learning something new about the world around us. And, for the most part, scientists are not driven by monetary gain. But if they discover something, they generally want the world to know it. Tantalizing visions of a Nobel Prize may influence their actions.

And, as I noted earlier, once human health is involved, the drive to find a new method or new drug can take on added urgency, for a medical discovery can be highly valuable not only to the cause of human health, but can provide professional (and monetary) advancement as well to the discoverer.

As a result, priority disputes in medical science can be particularly vicious. There are many examples in medical history that I looked into and could have chosen—examples include patent and priority battles among Morton, Wells, and Jackson in the development of a useful anesthetic; between Banting and Macleod in the discovery of insulin; between Guillemin and Schally in the discovery of the brain hormone; and between Gallo and Montagnier in the discovery of the AIDS virus. I chose the last one mainly because of its powerful, and continuing, relevance today. Curiously, as we'll see, these two both competed and cooperated.

The race to be first can also lead to superhuman effort, and, perhaps, even speed up the desired result. This seems to have been the case with Pasteur and his many battles (chapter 5).

A variation on the priority dispute involves arguments over whose method is better, as in the work of Sabin and Salk on a vaccine for poliomyelitis (chapter 8). Again, there are subtexts that are as interesting as the headlines. When Salk came up with the first real weapon against that dread scourge, he rapidly assumed the status of world-class hero. He suspected, however, that this meant trouble for him, and he was right.

In a different kind of dispute, two researchers may look at the same phenomenon and come up with different interpretations. This happened in the case of Galvani and Volta (chapter 2); the result was a lively controversy that dragged others in.

What, then, makes a feud great? From the many cases I scanned in making my choices, I have chosen those that seemed to me to have some special drama or scientific interest, that in some way influenced the future course of medical science, or that have had repercussions in our own day.

Freud (chapter 7), for example, is much in the news today. Freudians are celebrating the 100th anniversary of his pathbreaking *The Interpretation of Dreams.* But the controversy that still surrounds the very mention of his name began long ago. What we learn about his many battles with his contemporaries surely has relevance to the ever-present attempts to topple this icon, and perhaps can give those of us on the sidelines some greater insight into these attempts.

Another fascinating aspect of the Freud story is his connection with the medical establishment, again of current relevance because of the increasingly apparent linkage of mind and body.

In the case of Rosalind Franklin and Maurice Wilkins (chapter 9), two factors bring it into this book—the importance of the event, namely the discovery of the DNA double helix, and the poignant question of what might have been if the feud had not taken place. A case can be made that if Franklin had not feuded with her colleague Wilkins, she might have gone down in history as the discoverer of the double helix.

Here then is a personal look at medical history. In the same way that political history helps statesmen interpret today's events, so, too, may these dramatic cases help us understand the halting, confusing, but still wonderful world of medical research—and to realize that it is at least as much a human enterprise as it is an organized activity.

We begin our journey with the story of William Harvey and his ideas on the circulation of the blood. His courageous stand against powerful conservative forces provides an excellent start for our examination of some of the great feuds in medical science.

CHAPTER 1

Harvey versus Primrose, Riolan, and the Anatomists

Circulation of the Blood

In the 17th century it had been "known" for some 1,400 years that the blood was created in the liver, moved outward from the heart toward the extremities, and, in nourishing the tissues, just disappeared there. The heart was also the source of some sort of vital spirit, which in some mysterious way had to do with the blood.

In 1628, British physician and anatomist William Harvey announced his discovery of the circulation of the blood and, just as shocking, reported that the heart was merely a pump that was pushing the fluid around and around in the body.

He had laid out his theory, carefully and clearly, and of course in Latin, in a small book consisting of 72 poorly printed pages. Its title, *Exercitatio Anatomica de Motu Cordis et Sanguinis in Animalibus*,[1] is often shortened to *De Motu Cordis* (On the motion of the heart), or *DMC*. One reason for the poor print job was that it had to be published in far-off Germany, for the British censor forbad its publication, and no British publisher would touch it.[2] The publisher, Wilhelm Fitzer in Frankfurt, offered as an excuse for the many printer's errors the "unfavorable times," meaning the Thirty Years' War that was then ravaging Germany.[3]

Nevertheless, no one had any trouble figuring out Harvey's message. In the first half of the book, he presents his findings on the heart, and mentions some of his fears and hesitations. Though he published in 1628, he had started his work more than a dozen years earlier, and had been demonstrating parts of it at his own Royal College of Physicians for at least nine of those years,[4] hoping to build up support among his colleagues.

A quick look at the first chapter of *DMC* tells us what his experience had been up to the date of publication. His discovery, he writes, "pleased some more, others less; some chid and calumniated me and laid to me as a crime that I had dared to depart from the precepts and opinions of all anatomists."[5] He also lays out his research program in this first half and, mainly, gives the first scientifically based description of the heart's functions.

Then, in chapter 8, Harvey introduces the idea that the blood circulates, which, he feared, rightfully, would be even more shocking to his readers. In the opening paragraph, he states: "What remains to be said upon the quantity and source of the blood . . . is of a character so novel and unheard-of that I not only fear injury to myself from the envy of a few, but I tremble lest I have mankind at large for my enemies. . . ."[6]

Paranoia or Realistic Fears?

Was he just being paranoid? Hardly. Not long before, in 1591, Eufame Macalyane, a Scottish lady of rank, sought relief for the pain of childbirth. For this transgression—does not Holy Writ refer to "the primeval curse on woman"?—she was burned at the stake in Edinburgh.[7]

And in 1600, only a quarter century before Harvey's book appeared, Giordano Bruno, a pugnacious Italian philosopher, had been burned alive for his stubborn espousal of the idea that the universe was infinite and not bounded, as maintained by the Greek astronomer Ptolemy (second century A.D.). The Ptolemaic system had been strongly supported by the Church, which took Bruno's ideas to be sacrilegious and therefore dangerous.

A few years after Harvey's book was published, the great Galileo Galilei did recant when faced with the fearsome might of the Inquisition. As a result, he was merely sentenced to house arrest for the balance of his life—because he had published a book that argued against the same Ptolemaic system. But Bruno and Galileo were only the better-known examples of theological intolerance. Many others suffered and perished for even lesser crimes.

Granted, Harvey was an Englishman and an Anglican in a non-Catholic country. But England had been a Catholic country not long before and could be one again if the monarch chose to convert.

And Catholicism was not the only source of danger. In the middle of the previous century, Spanish-born Michael Servetus, a sort of itinerant physician with a strong interest in theology, had come up with a correct description of the pulmonary circulation (the movement of the blood from heart to lungs and back again). Unfortunately, his insight was connected with and contained within one of his many theological writings. This tract particularly *(On the Restitution of Christianity)* managed to aggravate both Protestants and Catholics, and, in 1553, it was burned by Calvin in Geneva—along with Servetus himself.

Witch-hunting and a widespread belief in the occult was another possible source of danger. In 1618, Harvey became one of King James I's personal physicians. But James was a strong supporter of witch-hunts, and had written a book on it. Harvey was fearful of the awful credulity of the time.

Harvey's era was also one of extreme political unrest. The Royalists, those supporting the king, and the Parliamentarians, looking to overthrow the monarchy, were constantly at odds. One never knew from which direction trouble might come. Harvey, as one of James's physicians, was blamed for bringing on the king's death (1625) when he defended the use of a remedy suggested by the king's favorite, Lord Buckingham. Fortunately, there were no serious consequences for Harvey, but the suspicions lingered.

All of this makes it a bit easier to understand why Harvey might be nervous, and perhaps explains why he walked around with a dagger strapped to his waist. The 17th-century biographer John Aubrey described him thus: "He was very Cholerique; and in his young days wore a dagger (as the fashion then was) but this Dr. would be apt to draw-out his dagger upon every slight occasion."[8] A more recent biographer, Geoffrey Keynes, adds that the dagger, which Harvey continued to wear in his middle years, reflected his earlier experience as a medical student at the University of Padua, where gangs of young students, representing factions from different countries, often battled each other.[9] There were also scuffles and even more serious disorders between students and townspeople.[10]

Harvey has been described as "rather on the small side, with raven hair, dark piercing eyes, somewhat sallow complexion, and a keen restless demeanour and rapid speech."[11] Science historian Jerome J. Bylebyl writes that Harvey "seems to have been well liked by those who knew him, although he was an outspoken man and perhaps somewhat short-tempered."[12]

A small, outspoken, somewhat short-tempered scientific genius in a superstitious and dangerous age. Is it any wonder that Harvey was nervous?

Initial Response

Harvey himself initially stood apart from the controversy that erupted after publication; but although he was not subject to physical violence, he took plenty of flak nonetheless. He was given the nickname Circulator, a conflation of the idea of circular reasoning with his theory of the blood's circulation.

The term's derivation also contains the implication of quack or mountebank. Aubrey wrote: "I have heard him say, that after his Booke of the *Circulation of the Blood* came out, that he fell mightily in his Practize, and that 'twas believed by the vulgar that he was crackbrained, and all the Physitians were against his Opinion. . . ."[13]

Viscount Conway advised his daughter-in-law not to use Harvey as her physician: "he is a most excelent Anatomist, and I conceive *[De Motu Cordis]* to be his Masterpiece . . . but in the practicke of Physicke I conceive him to be to[o] mutch, many times, governed by his Phantasy. . . . [And] to have a Physitian abound in phantasie is a very perilous thing. . . ."[14]

What Conway meant by "phantasie" was, clearly, Harvey's ideas about the circulation. Conway was writing in 1651, 23 years after *DMC* was published, by which time acceptance was relatively widespread. Yet the basic problem, that Harvey was attacking and overturning the ancient anatomists, still bothered many of his contemporaries.

The Galenic Corpus

Among those Harvey was overturning was Galen, a second-century Greek physician and anatomist. A genius and well ahead of his time, he produced enormous quantities of medical writings. Though Galen did not himself become a Christian, his search for design and purpose in all things, including the human body, made his works well-liked by the Church, and this in turn ensured the survival of his theories into Harvey's day.

Galen argued, for example, that nature always acts with perfect wisdom; and that the body is nothing but a vehicle for the soul. Christian theologians found that these ideas fitted nicely with some of their own: disease is punishment for sin, and the body is sacred. As a result, dissection of humans is a sin, and anything that detracts from the Great Physician is anathema.

Now, Galen had a high opinion of both himself and his work, but he recognized the likelihood of advances beyond his claims and explanations. He would probably have laughed had he returned in Harvey's day and seen what the academics had made of his doctrines—edifices to be admired, even revered, rather than foundations on which to build new ones.

One of these edifices was the *Anatomy* that Galen had put together, a comprehensive work that described much of the body and "explained" many of its functions as well. As attending surgeon at the gladiatorial games under the Romans, he had considerable experience with human bones, muscles, and blood. Not afraid to get his hands dirty, he had also done dissections and even experiments—which was more than could be said for most of Harvey's contemporaries.

There are hints that he did some of these on human bodies, but virtually all of his physiological research was on animals. He performed vast numbers of dissections and experiments on a variety of animals, from which he certainly learned a great deal about anatomy.

Unfortunately, he extrapolated what he learned from his experimental creatures to the human body. Some of it he got right, and much of it he got wrong. His descriptions of the muscles in humans were excellent, for example, and he did some good experimental work involving the spinal cord.[15]

But Galen's physiology—that is, his explanations of the functioning of the body's organs—comprised an amazing concoction of moonshine and faint glimmerings of nature's ways. An example of the moonshine was his use of "spirit." This vague concept was somehow connected with the blood. It also lent itself nicely to later religious and philosophical teachings involving the soul, the breath of life, and so on. Now, the air we breathe in and out does indeed have some sort of connection with the blood. But the spirit idea added nothing of value to anyone's understanding.

It could, however, be used in a vast variety of creative ways. As a Galenic follower explained in 1556: "During waking the face and other external parts are red, and well colored, each according to its nature; but they become pale and livid during sleep, which could only happen because at this time all the blood, or at least its lighter and more spiritous portion, betakes itself to the inner parts, while in waking it rushes out to the external parts."[16]

Fear was another example of the movement of heat, spirit, and blood to the inner regions of the body, while anger exemplified outward flow. Sadness was a less extreme inward movement, and joy an equally moderate outward dispersal. According to Galen, anger never caused any deaths, but some weak-spirited individuals may have died of overabundant joy. Sudden fear may also cause death, because the blood comes together and has a suffocating effect.[17]

By Harvey's day, medicine had solidified into a Galenic corpus. Teachers of medicine had relied for centuries on Galen's illustrations of the human body and saw no need to do any dissections themselves.

So religion, superstition, and medicine remained tied up in a huge Gordian knot. The Galenic corpus may be old; it may be wrong in many respects; but that doesn't mean it's simple. An edition of Galen's extant works, comprising 2.5 million words in 22 thick volumes, is estimated to be only two-thirds of his complete output.[18] Seventeenth-century anatomists and physicians didn't know a lot of medicine as we know it today, but that doesn't mean there wasn't plenty to teach.

Here I can only give the vaguest idea of the complex system of physiology they taught. The blood, said the Galenists, began in the liver, then moved outward from the heart and toward the extremities; it normally moved only in the veins and, in nourishing the tissues, just disappeared there, so there was no need for it to return anywhere.

The movement of blood was a volatile activity, not a regular thing, a response of some sort to a wide range of bodily requirements. It might be local in character, or it could be a massive movement both inward or outward. All of these were associated with a wide range of phenomena.

The arteries and the veins comprised two independent systems, each of which dealt with its own type of blood: venous blood had to do with nutrition, and the arterial with "vivification." The latter was a kind of pneumatic system that controlled the distribution of vital spirit and heat (including air) throughout the body.

Another source of trouble was Galen's belief that there are minute pores in the septum (the dividing wall between the two side-by-side ventricles of the heart), which permitted passage of substances, such as blood and spirit, between them. It was clear that there was some sort of connection between arteries and veins, so Galen invented the pores. He explained that it was not possible to see them "both because of their smallness and because when the animal is dead, all its parts are chilled and shrunken."[19]

Though hobbled by tradition and religion, however, 16th- and 17th-century surgeons, physicians, and anatomists still had eyes and brains. Experiments and observations did take place. It was their wonderful, creative explanations that kept the whole thing together.

They saw, for instance, that arteries and veins were physically different from each other. What then is the reason for the thicker, stronger construction of the arteries? That's easy. It is needed to hold in the strongly active and penetrating animal spirits.

The great anatomist Fabricius—Harvey's own teacher at the University of Padua—made what should have been a momentous discovery. In 1579 he described "little doors" (valves) in the veins. Fabricius, however, saw the valves as regulating the movement of blood to deal with pathological conditions. So he never understood their real purpose—permitting perfectly normal flow only toward the heart, exactly the reverse of what Galen had postulated.

Further, after a creature suffers a violent death, autopsy reveals that the veins are congested with blood, while the arteries are relatively empty. To the Galenists, this was proof that the arteries are normally filled with air and spirit, not blood. What actually happens is that the venous blood is held in place by the valves found everywhere in the veins, while the arterial blood has been free to flow out unimpeded.

Yet they also saw that if an artery is cut, it clearly spurts blood. This, however, was again taken to be a pathological condition: what is happening, they reasoned, is that as a direct result of the puncture, a sort of one-way evacuation is taking place, in which blood from the veins is drawn into the arteries and then expelled through the opening in the artery.

The worst part of it all is that these theories did not exist in a vacuum; they were drawn upon for clinical purposes. So if one is bleeding from an artery, the Galenists believed there must be too much blood in the veins, and that it has found its way to the arteries. Cure for puncture wound in an artery: bleed the veins.

In fact, all disease involved the distribution, or maldistribution, of blood and spirits, and it was relocation of these substances that they were trying to accomplish in their treatments. But Galen also supported Hippocrates' belief that disease resulted from an imbalance in the vital fluids, or humors, namely, blood, phlegm, black bile, and yellow bile. The easiest to get at, obviously, was the blood. This explains why bleeding, the deliberate letting of blood, played such an important part in Galenic medicine.

It's no wonder that Lewis Thomas could write that before the injection of science, medicine was an "unbelievably deplorable . . . story. . . . It is astounding that the profession survived so long, and got away with so much with so little outcry."[20]

Obviously, this couldn't change until a fresh look was taken at the physiology of the body, including the part actually played by the blood and, even more fundamentally, at that prime mover, the heart. In the Galenic view, the heart was the source of the vital spirit. It was also a flaccid organ; it swelled on a regular basis as a result of, rather than being a cause of, the movement of the blood. Until its proper function was understood, little more could be accomplished.

Steps along the Way

In the first chapter of Harvey's book he tells us:

> When I first gave my mind to vivisections, as a means of discovering the motions and uses of the heart, and sought to discover these from actual inspection . . . I found the task so truly arduous, so full of difficulties, that I was almost tempted to think . . . that the motion of the heart was only to be comprehended by God. . . .
>
> At length, by using greater and daily diligence and investigation, making frequent inspection of many and various animals, and collating numerous observations, I thought that I had attained to the truth. . . .[21]

A compulsive experimenter, he worked with all manner of creatures. But whereas Galen went off the track in many of his extrapolations from animals to humans, Harvey realized that he could actually learn more about the heart from some of the lower animals, particularly from cold-blooded creatures, because the movements of their hearts are slower and therefore easier to follow.

He points out in *DMC* that these movements "also become more distinct in warm-blooded animals, such as the dog and hog, if they be attentively noted when the heart begins to flag, to move more slowly . . ."[22] (meaning, of course, when the animals are somehow weakened or brought close to death). Over the long period of his theory's development, he may have experimented with some 80 species of animals.[23]

As he learned more about the heart, a huge puzzle presented itself. The traditional idea was that blood was formed from food in the liver; that it found its way through the system and out to all parts of the body; and that it was there consumed in the body's activities, including replacement of damaged tissue, as needed.

But Harvey, now convinced that the heart was the source of regular delivery, pursued a particularly unusual course for that time. First, he measured the volume of many left ventricles (the chamber that sends blood out to the extremities), and determined that in humans they contained an average of some 2 fluid ounces. Reasoning that even if only a quarter of the blood in such a chamber is sent out on each stroke, then the amount of blood being delivered—half an ounce times an average of 4,200 strokes an hour—would require the continuous production of over 65 quarts of blood per hour. This was clearly a preposterous idea.

The only logical explanation was that the blood was not being continually created and consumed, but was, to put it simply, conserved. Note that he was actually measuring something, in this case the volume of the left ventricle, and then putting the figure to work. In so doing he was creating a new world of experimental physiology. Of all his demonstrations and proofs, this one may have been the most telling in the long run.

But there were others. One problem he faced was that the optical equipment of his time was not sharp enough to detect the microscopic passages between the arteries and the veins in the extremities. But by a clever use of ligatures (straps) around the arm, by which he could selectively cut off and permit circulation in these vessels, he showed that blood did move from arteries to veins, though he could not say how. Whether other physiologists were convinced is another question.

But even with all this solid scientific activity, the *idea* of the blood moving in a circle may have come about in quite a different way, and again taps into the world in which Harvey lived and worked. The principle of circularity has a long history, and has been seen in a variety of applications, including the perfect orbital circles of the heavenly bodies in the Ptolemaic world system.

One of Harvey's colleagues, Robert Fludd, advocated a mystical system in the early 1620s that also involved the idea of circulation. The sun, he believed, issues some sort of "catholic spirit" (i.e., universal spirit), which gives life to the earth by penetrating a variety of "temples." These may be anything from the germ of wheat to the living heart, all of which behave like "suns." The pulsation of the heart, for example, distributes the vital spirit to the rest of the body by a process of circulatory currents, in the same way as the sun's catholic spirit spreads across the earth.[24]

Fludd, in other words, felt that both the Sun and the heart were generators of some sort and were involved in a kind of circulatory activity.

Of course, his circulation was quite different from Harvey's. But scientific ideas can arise in surprising ways.

Nor did Harvey simply leap out of the 17th century. Still an Aristotelian with metaphysical leanings, he wrote:

> The heart, consequently, is the beginning of life; the sun of the microcosm, even as the sun in his turn might well be designated the heart of the world; for it is the heart by whose virtue and pulse the blood is moved, perfected, and made nutrient, and is preserved from corruption and coagulation; it is the household divinity which, discharging its function, nourishes, cherishes, quickens the whole body. . . .[25]

We have no way of knowing for sure whether these metaphysical references reflected his own beliefs, or whether he included such writings to fend off attack.

Another interesting possibility: Jean Hamburger, a distinguished French medical researcher and writer, has recently suggested that the British political philosopher Thomas Hobbes may also have played some role in Harvey's vision of circularity. In an unusual kind of fictional biography, Hamburger postulates a diary entry by Harvey in which he writes: "While he [Hobbes] was thus evoking an image of a King whose mission it is to see that order and reason circulate among his subjects . . . another image was superimposed in my mind, that of a heart whose mission it was to circulate through the body the blood that bears the necessary spirits."[26]

Was Harvey influenced by Hobbes and his ideas about the state? They did meet sometime between 1621 and 1626. Although Hobbes's major works emerged well after Harvey had already come to his early conclusions, some early idea of Hobbes's might have provided a spark.

Interestingly, in Harvey's dedication to his royal patron, Charles I, he compares Charles to the sun. "The King, in like manner, is the foundation of the kingdom, the sun of the world around him, the heart of the republic, the fountain whence all power, all grace doth flow."[27] (Twenty years later the "foundation of the kingdom" was overthrown by the Parliamentarians and decapitated.)

One more possible influence on the development of Harvey's theory. Galileo was a professor at Padua when Harvey took his medical degree there. Galileo had been working on the mechanics of fluids in motion, including the theory of pumps. It seems reasonable to conclude that Harvey, exposed to this work, was influenced by it. Descartes, too, was working on a theory of motion at about this time, which may also have had some effect on Harvey.

These were some of the possible ideas and influences that got Harvey going. However it happened, Harvey came to his conclusions, built up a body of evidence, did what he could to convince his colleagues, and, finally, bit the bullet. The military metaphor is apt.

The Galenic Brigade

In the dozen years prior to the appearance of *DMC* in 1628, during which Harvey was developing his theory, preparing his manuscript, and lecturing on his ideas, there seems to have been little argument against his work, at least not in print. Harvey's reputation remained secure. His friend Fludd even published a work shortly after *DMC* appeared that supported Harvey. It was, however, mainly a metaphysical treatise, with little scientific discussion of Harvey's ideas. In fact, his main point was that Harvey's work provided support for his own mystical ideas.

There were a few other halfhearted efforts at support. Descartes issued the first actual discussion—meaning with some scientific content—of Harvey's work in 1637, but argued against his theories on the heart.[28] The earliest real, published support among Englishmen seems not to have appeared until 1641 (George Ent) and 1644 (Kenelm Digby)—which must have seemed like forever to Harvey. Digby, in fact, was finally answering Descartes's objections to parts of the new theory.[29]

On the other hand, three book-length refutations appeared in short order. The first, in 1630, came from James Primrose, an English provincial physician and a voluminous writer.

His work, with a title even longer than Harvey's[30] (but ending in *adversus Guilielmum Harveum*), was a compilation of misreadings, misunderstandings, and misstatements, at least to a modern reader. Robert Willis, the translator of *DMC* whose work I am using, wrote in 1878 that Primrose's book consisted of "obstinate denials, sometimes of what may be called perversions of statements involving matters of fact, and in its whole course appeals not once to experiment as a means of investigation."[31]

Primrose's style, however, is interesting. Here and there he addresses Harvey directly in apparently worshipful tone:

> Thou has observed a sort of pulsatile heart in slugs, flies, bees, and even in squill-fish [a kind of shrimp]. We congratulate thee upon thy zeal. May God preserve thee in such perspicacious ways. . . . Those who mark in thy writings the names of so many

and diverse animals will take thee for the sovereign investigator of nature and will believe thee to be an oracle seated upon the tripod. . . .

Then comes the snide attack:

I speak of those who are not physicians and have but a smattering of the science. But if we read the works of real anatomists, such as Galen, Vesalius, the illustrious Fabricius and Casserius, we see that they have provided us with engraved plates representing the animals they have dissected. As for Aristotle, he made observations of all things and no one should dare contest his conclusions.[32]

Primrose added that he had managed to refute in 2 weeks what it had taken Harvey 20 years to develop.

Primrose was generally contentious; not only did he have difficulty accepting new ideas, but he seemed compelled to fight them in print. More than a dozen years later, with the circulation idea gaining adherents, he was still in action. In 1644 he engaged in a pamphlet war with a defender of blood circulation.[33]

Primrose's contentiousness could be part of the reason for his attack on Harvey. He had also, however, made application to the College of Physicians in December 1629–January 1630, and was not accepted; he may have put at least part of the blame on Harvey for his rejection and was probably angry about Harvey's continued high position at the college in spite of his crazy anti-Galenic ideas.

But the plot may be even thicker than that. There is evidence that the president of the college, Dr. John Argent, exhorted Primrose to mount his attack.[34] Yet Harvey considered Argent a good friend and even dedicated his book to him.

In 1632 a respected Danish professor of medicine, Ole Worm,[35] also looked into the circulation question. He apparently teetered for a bit, but when he balanced Harvey's evidence against the strength of the old masters, Harvey came out second best. Worm let loose the second major critique of the new idea.

Then, in 1635, Emilio Parigiano, an elderly and respected Venetian physician, gave the world yet another extensive critique in the form of lengthy abstracts from *DMC* with, on the opposite page, acerbic criticism of each section. Regarding specific heart sounds, for example, which Harvey considered important, he suggested that Harvey was a victim of his own imagination. "Our poor deaf ears, nor those of any physician in Venice, cannot hear them; thrice fortunate those in London who can."[36] He also, says historian Roger French, "saw Harvey as attempting to overturn the rationality and providence not only of nature, but of God."[37] Parigiano, interestingly, had studied with the eminent Fabricius, the same anatomy teacher Harvey had studied with in Padua.

In 1636, Harvey was in Nuremberg and paid a visit to the University of Altdorf, where he gave a public demonstration of his new theory. The main objective was to win over a distinguished anatomist and professor of medicine, Caspar Hofmann, another implacable foe of the new idea. Hofmann attended and promised to give Harvey his answer the next day. He did, in the form of a letter.

It began: "Your unbelievable kindness, my Harvey, makes me not only like you but love you . . ." After more of the same, he changes course: "You appear to accuse Nature of folly in that she went astray in a work of almost prime importance, namely, the making and distribution of food. Once that is admitted, what degree of confusion will not follow in other works which depend on the blood?"[38] He promises, however, that "if, after the clouds have been dispelled, you will show me the truth which is more beautiful than the evening and the morning stars, I will . . . publicly recant and retire from the field."[39]

Harvey, perhaps believing Hofmann's promise, replied immediately with a long, carefully crafted letter. His manner, as demanded by the times, is as courteous as Hofmann's, but he used some fairly direct words: "First you thought fit to indict me . . . because I seemed to you to charge and convict Nature of folly and of error, and to characterize her as a very stupid and idle worker . . . But . . . as I have always been full of admiration for Nature's skill, wisdom, and industry, I was not a little upset to have been given such a reputation by a man so very fair-minded as yourself."[40] He then goes on to answer Hofmann's objections. They corresponded further and Harvey even visited again. All to no avail.

Later years brought a truly strange twist of the knife. Harvey began to hear of lecturers on anatomy in England—where acceptance grew much faster than elsewhere—who were giving credit for the discovery of the circulation of the blood to Hofmann! In one case Harvey actually had to compare letters and dates in order to defend and clear himself.[41]

Among the many attacks on Harvey was one that could have been dreamed up by a spymaster like John Le Carré. In 1639 there had appeared a pamphlet titled *A Most Certaine and True Relation of a Strange Monster or Serpent Found in the Left Ventricle of the Heart of John Pennant, Gentleman, of the Age of 21 years,* by Dr. Edward May, physician to Queen Henrietta Maria (wife of Charles). In it May details the results of an autopsy that showed clearly a "worme or serpent" coiled up in the cavity of the young man's left ventricle.

This was, ipso facto, a kind of proof that Harvey was wrong; a monster living in the ventricle showed that the blood does not circulate (if it did, the serpent would have been flushed out). Bad enough; but then, in 1643, Marco Aurelio Severino, a distinguished professor of anatomy at Naples, published the second edition of his important textbook *De Recondita Abscessuum Natura* (Of the hidden nature of abscesses), and included an account of May's finding.

But Severino, located in Naples, had actually written to an English colleague, Dr. John Houghton, asking for some details of the monster. Houghton had shown the letter to Harvey, who had quickly understood what was going on. May had actually found a plain old blood clot, but one in the form of a "serpent," which had slipped into the heart from a nearby blood vessel.

This is a very puzzling story, however. In an exchange of letters, Houghton must have told Severino about Harvey's conclusion. Yet although Severino is effusive about how much he admires "the great Harvey, that pillar of England as well as medicine and anatomy,"[42] he nevertheless included May's description of the "cardiac worm" in his textbook.

The "finding" was again recounted in a 1678 book called *Wonders of the Little World,* and apparently remained in the literature for another century.

It was also put to use by one of Harvey's most vicious enemies, though perhaps unwittingly. In 1648, a full two decades after publication of *DMC,* Jean Riolan, dean of the Paris Faculty of Medicine and personal physician to both the French king and the queen mother (and one of Primrose's teachers in Paris), unleashed the first of two major attacks on Harvey's work. The first (1649) was relatively gentle and attacked him only indirectly.

The second, *Opuscula Anatomica Nova,* came a year later; it is much more vituperative and even goes after a Leyden physician, Jan de Wale, who had supported Harvey's ideas. This, in spite of the fact that de Wale claimed the idea had its origin in antiquity and that other workers had brought it along to the point where Harvey could, and did, merely confirm the theory.[43] Harvey faced this problem over and over.

Riolan's attack reiterated the old arguments; for example, "Harvey is very learned, but when he says that the blood passes through the lungs, he is going against Nature."[44]

Now, Riolan was neither a fool nor a mountebank. He recognized that there were some discrepancies between Harvey's (and others') new findings and those of Galen. He felt, however, that when experimental findings contradicted Galen there must be something wrong with the new findings. Experiments might also create experimental injury that would destroy the physiological conditions and prevent accurate observation.[45] Other possibilities included a deterioration of the Galenic texts; minor errors of Galen owing to his lack of human cadavers; and changes in human physiology since Galen's time due to the influence of climate, soil, and diet.[46]

He felt that there was some circulation of the blood going on, but not as Harvey saw it. "I assert," he wrote, "that the use of the circulation lies in the uninterrupted generation of vital blood and the maintenance of a continual heart beat."[47] One of his major objections was that if Harvey was right, the liver lost its central position as the source of the blood.

What made Riolan's objections so worrisome was his high standing in the world of medicine; his skill as an anatomist was celebrated throughout Europe. In fact, anatomy students still read today of Riolan's arch (in the colon), Riolan's bone (a small bone at the back of the head), Riolan's nosegay (a small group of muscles in the same region of the head), and Riolan's muscle (in the eyelids).

Riolan hoped by the strength of both his reputation and his "arguments" to finally put Harvey's nasty theory out to pasture. What must have been particularly galling to Harvey was that he had actually cited "the learned Riolanus" in the introduction to *DMC*.[48]

Through all of the preceding two decades Harvey had shown enormous self-restraint, hoping that his colleagues would step in to counter the continuing tirades. Unfortunately, there was precious little of such support.

Some of Harvey's reticence had to do with his own reluctance to engage in controversy, and some to the continuing problems that plagued the England of his era. Part of the time he spent in exile with the king; in the early 1640s the Parliamentarians ransacked his apartment and made off with some of his notes and manuscripts. As a member of the College of Physicians, he also had to deal with opposition from two other groups looking to assert their independence and authority: the apothecaries and the surgeons.

Although Harvey was reluctant to engage his opponents in print, he had no compunctions about complaining to others, and particularly about Riolan. When a German colleague, Johann Daniel Horst, wrote to Harvey asking about Riolan's claims, Harvey answered:

> [Riolan] has very obviously achieved mighty trifles with great effort and I cannot see that his fictions have brought pleasure to anyone. Schlegel[49] wrote more carefully and modestly and, had the fates permitted, would doubtless have taken the force out of Riolan's arguments and even out of his taunts. But I learn, and that with sorrow, that he shuffled off this mortal coil of ours a few months since.[50]

Harvey Responds to the Attacks

Finally, Harvey could stand no more and, in 1649, he issued the only published response he ever made to criticisms of *DMC*. More than two decades had passed, an astonishing example of self-control. He penned two essays in the form of letters to Riolan, a fairly common form of scientific communication.

In published form, his essays/letters are generally known as *De Circulatione*.[51] The first letter, probably written in 1648 or 1649, presents a set of detailed answers to Riolan's objections. The first three paragraphs and the closing paragraph of the second letter also direct answers to Riolan. But the rest answers many of the objections made by one and all in the years leading up to this point, and includes a description of four new experiments he had performed since publication of *DMC*.[52]

Like the interchanges mentioned earlier, the second letter starts off with some courteous, even courtly, language. But it also includes some honest venting of feelings long pent up against "those who cry out that I have striven after the empty glory of vivisections, and [who] disparage and ridicule with childish levity the frogs, snakes, flies, and other lower animals which I have brought on to my stage. Nor do they abstain from scurrilous language."

Harvey writes that he refuses to "return scurrility with scurrility." Nevertheless, he adds, "It is unavoidable that dogs bark and vomit their surfeit . . . but one must take care that they do not bite, or kill with their savage madness, or gnaw with a canine tooth the very bones and foundations of truth." Finally, "Let them enjoy their evil nature. . . . Let them continue with their scurrility until it irks if it does not shame them, and finally tires them out."[53]

He kept to his word. His only other major publication, *De genera-tione* (1651), summarized later work he had been doing on embryology. True, his intent in these efforts was, at least partly, to buttress his work on circulation, but he steered clear of the controversy itself.

Because of Harvey's early experience, this publication had to be pulled from him by the same Dr. Ent who had been one of his early supporters. Ent later reported that Harvey argued at first. "Do you really wish once again to send me out into the treacherous sea away from the peace of this haven [his brother's house] in which I pass my life? You know well how much trouble my earlier studies evoked. . . ."[54]

Nevertheless, Harvey was by this time enjoying a rather different atmosphere than he had faced a quarter century earlier. Though there were still objections being voiced, and complaints about the overturning of ancient wisdom, he had managed to live long enough to see his theories widely accepted and even admired. Science historian I. B. Cohen maintains that Thomas Hobbes's seminal masterpiece, *Leviathan* (1651), was heavily influenced by Harvey's work.[55] Harvey seems also to have found a kind of peace in his researches, both earlier and later in his career.

We don't hear much of his wife, but she was apparently a source of strength and quiet pleasure over the 40 years of their marriage. She died in 1646, after which he spent his later years living with his brother, Eliab.

In that same period he was offered the presidency of the college he loved for so long and so well. Pleading ill health and the weariness of old age, he turned it down. He appears also to have turned down the offer of certain titles, referring to them as "wooden leggs," which was his way of showing his valuation of learning over rank.[56]

Abraham Cowley, in his "Ode, Upon Dr. Harvey" (1663),[57] refers to:

Coy Nature, (which remain'd, though Aged grown,
A Beauteous virgin still, injoyed by none,
Nor seen unveiled by any one)
When Harvey's violent passion she did see,
Began to tremble and to flee . . .

But, says Cowley, eventually she would strike back:

For though his Wit the force of Age withstand,
His Body alas! and Time it must command,
And Nature now, so long by him surpass't,
Will sure have her revenge on him at last.

But not until 1657, at which time Harvey had reached the age of 79, a small miracle in those days, and with faculties intact and active till the last. He had become a man of wealth, but had simple tastes and was generous both publicly and to his friends and family.

Bishop Brian Duppa, a friend and neighbor of Harvey in his later days, considered Harvey a good example of long life, and felt that "being of a dry sear body, he praeserved it so long by the rules of art and diet . . . Then," Duppa adds, "his life went out like a spark, without any violence or noise at all." [58]

Though Harvey continued to express some bitterness and cynicism in his later years, the consensus seems to be that he managed to transcend the difficult years and to show an essential humanity and to exhibit, as one of his colleagues put it, a kind of "facetious [witty] courteousness." [59]

One of Harvey's problems was that he could never explain how the interchange occurred between venous and arterial blood. In the same year that *DMC* appeared, there was born the infant who—33 years later, and 4 years after Harvey's death—provided the proof Harvey had lacked. Using an improved microscope, Marcello Malpighi, an Italian physiologist, was able to show that arterial blood does not simply leak out into the tissues, to be collected somehow by the veins, as was commonly believed. Rather, his microscopical studies showed clearly the exquisitely tiny capillaries; these, distributed throughout the body, connect the two sets of vessels. The circuit was, finally, complete, and Harvey could rest in peace.

CHAPTER 2

Galvani versus Volta

Animal Electricity

Why would Luigi Galvani, Italian physician and anatomist, have an electrical generating machine in his laboratory? Although the 20th century would see the development of electroshock machines for depression, electrocardiographs and pacemakers for heart problems, and a myriad of other electrically based medical equipment, there was none of that in 1790, when Galvani was in his heyday.

Still, devices for generating and storing electrical charge had already been developed, and were still new enough to fascinate anyone who could put hands on them. A rotating friction machine, such as the one Galvani used, could generate charge on a continuous basis.[1] The Leyden jar,* which could store up enough electricity to deliver powerful shocks, was another common piece of electrical equipment. Both of these are still commonly used in students' physics laboratories and in lecture demonstrations.

If one of the participants was touching a rotating charge generator, the act of kissing permitted the built-up charge to discharge to ground via his or her lips. Electric kisses became all the rage.

The idea of using electricity in medicine also took root quickly. Newspapers and rumor mills were filled with reports that electricity had been used to cure an astonishing range of maladies, from constipation to paralysis, from headache to herpes.

"By the 1780s," Marcello Pera writes, with tongue firmly in cheek, "electricity had truly become a panacea and even a miracle, as in the case of the couple who, after ten years of infertility, 'regained hope' through electricity–in [Abbé] Bertholon's words–thanks to a few

---

* The Leyden jar is basically two conductors separated by a nonconductor. It can build up a powerful charge.

turns of the crank and some shocks in the appropriate parts (the Abbé, demurely, did not specify which parts)."[2]

It wasn't all foolishness. After all, experimenters saw that electricity could have a powerful stimulatory effect on pulse and respiration. Earlier, in 1615, Nicholas Godinho had observed that if a newly caught electric torpedo (a marine animal that can serve up powerful shocks) was thrown into a pile of dead fish, some of them seemed to come to life.[3] By the middle of the 18th century, physiologists had shown that discharging electricity into muscle could cause paralyzed limbs to move. This led to the tantalizing suspicion that electrical stimulation might somehow be useful in restoring mobility to damaged or paralyzed limbs.[4]

In all cases, the results were tantalizing but, for lack of any real understanding of what was happening, inconclusive. Between 1750 and 1780, some 26 major publications appeared in France alone describing attempts to use electricity to create movement in paralytic patients.[5]

Controlling use of paralyzed limbs using electrical current remains today a hope and an active field of research.

Animal Electricity

Searching for background material on this chapter I came upon a book, published in 1960, that contained the following story.

> On an evening in 1780 Aloisio Galvani,[6] Professor of Medicine, was lecturing to his pupils at his home in Bologna. . . . Signora Galvani was sitting at her accustomed place at the hearth, eagerly listening, as always, to her husband's lecture. She had been using . . . the surgeon's knife with which her husband made anatomical dissections . . . to skin frogs in order to cook their thighs for supper. . . .

The story goes on to tell how the steel knife fell from her hand onto an exposed nerve of a frog's thigh, and at the same time touched the tin plate on which it rested. At that moment a spasm ran through the skinned thigh, and it stretched as if to jump away.

"Her first impulse," the author continues,

> was to stab the frog, which she supposed to be still alive. But the frog's legs relaxed and moved no more. Signora Galvani now touched it again in the same way as before, and again it twitched.

To this intelligent and observant woman it was at once clear that this must be some hitherto unknown phenomenon in natural science. She drew her husband's attention to it, and Galvani exclaimed:

"Wife, I have made a great discovery—animal electricity, the primary source of life!"

Later in the chapter, the author complained that Galvani, in his writings, "made no mention at all of his wife's part in the matter. . . ."[7]

Great story, I thought, but it flew in the face of everything I had read earlier about Galvani. The author, Egon Larsen, was a Briton about whom I knew nothing. I knew that I couldn't use it unless I could find some confirmation. I searched further and found just what I was looking for—the same story in a different book. Unfortunately it had also been written by Larsen.[8]

To compound the "Larseny" (stealing credit from Galvani), Larsen also wrote, "The clever woman had made an important discovery, and her not very intelligent husband did not know how to set to work on it. . . ."[9]

Did Larsen know something that other writers in the field had missed?

Inspiration and Influence

On the one hand, the level of Larsen's scholarship is indicated by an important omission: He neglected to mention that the contractions in these early experiments only occurred when a device for generating static electricity[10] was in operation nearby. As we'll see later, this machine played a crucial role in the discovery story.

On the other hand, there may be a germ of truth in Larsen's story. Galvani does *not* mention his wife in his summary of his experiments, published in 1791. Yet there are historical hints that she did play some part.

Pera, who has done what is probably the definitive biography of the controversy, points out that Galvani's wife played a "major role in his spiritual and scientific life."[11] She was also the daughter of Domenico Gusmano Galeazzi, a physicist and anatomist who had at one time been Galvani's teacher.

In an 1881 book on the history of science, Arabella B. Buckley wrote: "This discovery was made by Galvani . . . , or perhaps we ought to say by Madame Galvani, for it was her observation which first led her husband to study the subject." In fact, Buckley's version of the story parallels Larsen's, except that she includes the all-important electrical machine.[12]

Is Buckley's version an example of feminist revisionism a century before its time?

No. Professors J. F. Fulton and H. Cushing—both decidedly male—summed up the question nicely in an important bibliographical study of the early writings: "Whether Lucia Galeazzi [Galvani's wife] was an active rather than a passive agent in the making of the primary observation it is now impossible to say, and the usual story of Galvani having skinned some frog legs to prepare a bouillon for her during an illness may be wholly apocryphal."[13]

One problem with the bouillon idea is that it makes the whole discovery sound like pure accident. But this is far from the truth. Pera points out that Galvani began working with frogs' legs in the latter half of the 1770s, observing the contractions of animal muscle under both mechanical and electrical stimulation.[14]

And if it wasn't Galvani who actually made the first observation, no matter; he recognized that something significant was going on, and performed hundreds, perhaps thousands, of experiments and observations before he published his work.

Whatever the truth of the discovery story, we know from some early writings of Galvani's that he was aware of the electrical/medical experiments and, from his own researches, was pretty well set on the idea of animal electricity by 1786.[15] But no one paid much attention until his 1791 publication.

De Viribus

Galvani's account of his work is titled *De Viribus Electricitatis in Motu Musculari Commentarius* (On the effect of electricity on the motion of muscles). In it, he spells out a careful, sensible set of experiments and observations that show the development of his idea of animal electricity. This is hardly the work of a "not very intelligent" man. Rather, we see a serious, dedicated medical researcher wrestling long and hard with a chance discovery in a highly confusing field, one that led eventually to the complex world of electrophysiology—the scientific study of the part played by electricity in the activities of the body. Such activities are still complex and, in some cases, still confusing.

At the time, however, much of medicine still looked to animal spirits and other "subtle fluids" as a useful explanation for the body's activities and for a variety of ills as well. It was common to think of this fluid as flowing into the muscles from the nerves in living animals, causing their movement.

Galvani went along with this idea, but added much more to it.

Second Experiment

It was common in his day to prepare frogs for dissection by hanging them outside on metal skewers. This might have given Galvani his next idea. He wondered whether the muscle contraction effect would take place in the open air during thunderstorm activity. Benjamin Franklin and others had already shown that lightning was a powerful discharge of static electricity, and was therefore the same sort of electricity as was produced by down-to-earth electrical devices.

He set up an experiment in his backyard that re-created the original one; in this case the prepared frog formed part of a circuit that involved two different metals—brass hooks and the iron fence in his yard. Had the fence been made of wood the story might have ended differently.

But he did observe contractions in these circumstances. Then, good researcher that he was, Galvani tried the same thing in calm weather, when there were no storm clouds in the area. In his words:

> I had upon occasion remarked that prepared frogs, which were fastened by brass hooks in their spinal cord to an iron railing which surrounded a certain hanging garden of my home, fell into the usual contractions not only when lightning flashed but even when the sky was quiet and serene. I [therefore] surmised that these contractions had their origin in changes which occur during the day in the electricity of the atmosphere.
>
> Hence, with confidence I began diligently to investigate the effects of these atmospheric changes on the muscular movements I witnessed and I repeated the experiment in various ways. Therefore at different hours and for a span of many days I observed the animals which were appropriately arranged for this purpose, but scarcely any motion was evident in their muscles. I finally became tired of waiting in vain and began to press and squeeze the brass hooks . . . against the iron railing [with the result that] I did observe frequent contractions, but they had no relation to the changes in the electrical state of the atmosphere.[16]

He was tempted to believe that the contractions resulted from atmospheric electricity that slowly insinuated itself into the animal, accumulated there, and then was rapidly discharged when the hook come into contact with the iron railing.

But, recognizing that "it is easy to be deceived and to think we have seen and detected things which we wish to see and detect," he tried the same experiment indoors and,

> behold, the same contractions and movements occurred as before. I immediately repeated the experiment in different places with different metals and at different hours of the day. The results were the same except that the contractions varied with the metals used. . . . Then it occurred to me to experiment with other substances that were either non-conductors or very poor conductors of electricity. . . . Nothing of the kind happened and no muscular contractions or movements were evident. These results surprised us greatly and led us to suspect that the electricity was inherent in the animal itself.

He added that the observation of "a kind of circuit of a delicate nerve fluid . . . from the nerves to the muscles when the phenomenon of contractions is produced . . . strengthened this suspicion and our surprise."

He had, in other words, observed that contractions can be caused in a dissected frog by connecting the exposed nerve and muscle together via a metallic "arc." This was his famous "second experiment."

He believed that he had shown conclusively that the electricity was inherent in the animal itself, and he named it animal electricity. He felt, too, that he had finally solved the puzzle of the animal spirits with which medical people had been wrestling since at least the time of Galen. Even the mysterious operations of the nervous system seemed clearer now.

As he understood the situation, the animals were merely reservoirs of two kinds of electricity. As in a Leyden jar, these two kinds were separate and were stored, but in the frog's case they were positive in the nerves and negative in the muscles. He saw the metallic connectors being simply the conductor that discharges the difference between the two reservoirs.

Careful experimenter that he was, he had actually noted that the effect was stronger when two metals were used as a connector, as with the brass hooks against the iron railing, even when no electrical machines were operating nearby. He had also seen that certain metals worked better than others.

And right here the entire future history of electricity—the absolute bedrock of our modern civilization—lay exposed to him. Right here lay the chance for him to go down as the superhero of his day, the discoverer of current electricity as well as of animal electricity.

But Galvani was a physician and an anatomist, not a physicist. And rather than concentrating on, and following up with, more experiments with the metals, he chose the other possible route—animal electricity. It was a decision that was to have serious consequences, for science and for him.

A simple description of what happened here is that Galvani had made an important discovery, but failed to understand it. Then someone else picked up the idea and ran with it, scored the winning touchdown, and went on to everlasting glory.

The truth is far more complicated.

And on This Side of the Ring, Alessandro Volta

The immediate consequence of Galvani's publication of *De viribus* was an explosion of interest and activity. The biomedical world ran to its electrical machines; physicians, physicists, and laymen alike redid his second experiment—it was, after all, fairly simple to do.

Among those who got involved was Alessandro Volta, a highly respected physicist. In the same year as Galvani published his *De viribus*, Volta had been elected a fellow of the prestigious Royal Society of London for his contributions to the emerging field of electricity. Most of the physical research in those days consisted of learning how to detect, store up, and measure static electrical charges, and one of Volta's major discoveries was a highly sensitive device for detecting electric charge.

So it's not surprising that Galvani sent a copy of his *De viribus* to Volta, no doubt looking for and expecting his approbation, which he got—at first. Volta, in his early response, referred to "the fine and grand discovery of an animal electricity, properly so called. . . ." [17]

It didn't take long, however, before Volta began to insert his own spin on the observations. He carried out his own set of experiments and began to feel that Galvani had misunderstood what he saw, that there was no such thing as animal electricity. Volta showed that muscles could be made to contract even when they were not part of an electrical circuit, that they would contract even when metals were merely touched to the nerve.

As he put it, had Galvani "a little more varied the experiments, as I have done . . . , he would have seen that this double contact of the nerve and muscle, this imaginary circuit, is not always necessary. He would have found, as I have done, that we can excite the same convulsions and motions in the legs, and the other members of animals, by metallic touchings. . . ."

He concludes: "It is thus . . . that I have discovered a new law, which is not so much a law of animal electricity, as a law of common electricity, to which ought to be attributed most of the phenomena, which [although they appear] to belong to a true spontaneous animal electricity . . . are really the effects of a very weak artificial electricity." [18]

In other words the electricity was not of animal origin, but resulted from the metals themselves—that is, from the contact between the two metals used in Galvani's second experiment and his own.

As I describe the experiments, remember that the ones I mention have been culled over the years as the important ones. At the time, however, their significance was far less obvious. Giorgio de Santillana, a well-known science historian, refers to the "fearfully intricate situation with which the experimenters were confronted." [19] Many scientists, in fact, criticized both Galvani and Volta. [20]

Nevertheless, a true tug-of-war developed between the two researchers and those who followed them. Both men knew something important was in process, and each small observation brought an almost immediate refutation and a different angle.

For example, Galvani's team responded to Volta's claim by showing that the effect took place even if two metals of the same material were used in the discharge process. Again, this looked like the end of Volta's idea.

Not so, countered Volta; the metals may look the same, but may very well be different in some way—in their temperature, in the annealing process when they were made, or even from some minor difference in the polish.

In response, Galvani's side produced muscular contractions by simply folding the frog's exposed nerve back onto the muscle, without using any metal whatever—Galvani's so-called third experiment.

Galvani Moves Ahead

In the early years of their conflict, things looked good for Galvani. In 1794, for instance, the eminent naturalist and biologist Lazzaro Spallanzani wrote in a letter to a colleague:

> Today, Alessandro Volta, awarding degrees to some Engineers, read a long, long speech, entirely directed against Signor Galvani's electricity. In it, he claimed to prove that this electricity should not be properly called *animal,* but rather *metallic,* as he considers it to be exclusively generated by the armatures [conductors]. He based himself on several of his experiments, which, as is his wont, were awash in a sea of words. But for us—his colleagues—who listened to him, he failed to dispel our factually-based opinions in favor of a truly animal electricity.[21]

Volta, on the other hand, also expressed himself freely in his letters. In one he wrote, "I know those gentlemen want me dead, but I'll be damned if I'll oblige them."[22] It was also clear that he was deeply wounded by what the 19th-century French physicist Dominique Arago later described as "the tone of assurance with which . . . the galvanists, *old and young,* boasted of having reduced him to silence." Arago added, however, that this silence "was not of long duration."[23]

Volta countered Galvani's third experiment by arguing that it only showed that his law had a broader implication: practically any two dissimilar substances, whatever their nature, create electricity by their very contact, though the effect is stronger in some materials than in others.

And so it went, back and forth, with participation, on both sides, spreading to France, England, and Germany. But then two highly significant events took place.

As I noted earlier, Galvani's wife had played an important part in both his personal and scientific life. When she died in 1790 he took it hard. Then, in 1797, Napoléon's empire building took another severe personal toll on him. For Bologna, formerly a Papal State and Galvani's base, became part of the new Cisalpine Republic created by Napoléon. Galvani was called upon to take a civil oath swearing allegiance to the republic. He refused, and was immediately cut off from his professorship, his academic privileges, and his income. With that, he lost whatever leverage he might have had to help him in his continuing battle with Volta and Volta's followers.

Galvani found all of this very painful. Although the administrator of the republic eventually restored both his position and income, Galvani died in 1798, before he could resume his position.

Volta

At almost the same time Volta, seeking to find a way to demonstrate the truth of his theory, came up with a blockbuster of a demonstration, which he announced in 1800. He put together pairs of dissimilar metal disks, each disk separated from its adjoining one by a moist, nonmetallic conductor, and with each pair electrically connected to the next pair. Placing them one on top of another, he came up with what he called his electric pile. If the circuit was routed through an electrical measuring device it was clear that—for the first time in history—a continuous electric current was being produced. It was the predecessor to the modern battery and a major turning point in the scientific revolution that was to follow.[24]

Ironically, while Galvani had compared his animal electricity to the physicist's Leyden jar, Volta compared his voltaic pile to the biologist's electric torpedo; for dissection had shown that these creatures also contain a series of electric organs connected one to another.

As in the torpedo, the more disk pairs used, the greater the electrical current the pile produced. As occurred at first with Galvani's discovery, there was a great uproar. The early experiments and demonstrations were fairly unfocused, and looked more to the newfound power of the electric pile. Experimenters strung together hundreds of disk pairs and produced powerful shocks. Physicians may have used the numbing effect of powerful shocks for pain, headache, and gout, as Scribonius Largus did with electric fish in Rome during the first century A.D.[25]

Quickly, however, other aspects of the discovery started to emerge. The British chemist Humphry Davy used electrical power to split substances apart and thereby discovered several elements, including sodium and potassium. He used the power of the pile to build an electric arc lamp, the first true electric light. He even designed an electrolytic process for desalinating seawater. And all within a decade of Volta's discovery.

Volta's star rose rapidly. Whereas Galvani suffered from Napoléon's activities, Volta, more able to sway with the political currents, was accorded high honors by that fickle leader.

But the Galvanists' fires had been stoked long enough that they were not about to go out just like that. Galvani, though dead for several years, still had his followers, especially those in Bologna who were quite happy to continue the battle with their traditional rivals at Pavia. But now they made a mistake. Pera explains:

> When you are engaged in a dispute a wrong move may be fatal. Greatly surprised by the pile, Galvani's followers tried to confront Volta's ideas by introducing a hypothesis that postulated the existence of either (1) two distinct electrical fluids, the common and the galvanic, or (2) a single fluid that, in animals, underwent some change and acquired new, special properties. As the hypothesis proved untenable, its collapse automatically gave the edge to Volta's ideas.[26]

This, says Pera, was the main reason Galvani's side lost one of the major scientific battles of all time, and lost him the chance to go down as the original discoverer of current electricity.

But there was more. As the argument began to tilt in Volta's favor, none of the Galvanists seemed up to a very difficult task. And, in truth, there simply was not enough known at the time to see what was actually happening. Of those who did take up arms—rhetorical arms, that is—the main one was Galvani's nephew Giovanni Aldini

Aldini, Scientist and Showman

Aldini, the son of Galvani's sister Caterina, had had a close relationship with Galvani and even worked with him on many of his experiments. In fact, Volta had early on begun to direct his communications not to the shy and retiring Galvani but to Aldini, who had become professor of physics at the University of Bologna. Aldini sponsored a society at the university with the professed purpose of combating Volta and the group established at the University of Pavia who provided support for Volta.[27] Spallanzani and others expressed admiration for some of Aldini's early work.[28]

After Galvani's death, however, Aldini lost his moorings and began to thrash about. Though his objective, to defend Galvani and his ideas, was noble, his methods and approach probably did more harm than good.

Changing the action from scientific research to the realm of public entertainment, he went off in some wild directions, which included some bizarre spectacles. In 1803 he gave public demonstrations at several London hospitals. His most dramatic performances included zapping the heads of decapitated animals and even the beheaded corpse of a just executed murderer. Using a voltaic pile of 120 copper and zinc pairs, he caused eyes and tongues and other parts of the body to move and jerk about. Unhappily, and as a result, Galvani's work became associated with circuslike entertainment and with a variety of medical treatments that ranged from the absurd to the ridiculous.

Though electromedicine was only one of many off-the-wall methods sweeping the wide world of illness and health, it was an important one if only because of its great promise—and great mystery. In the same year as Aldini was performing in London, a medical group started a journal devoted to articles dealing exclusively with galvanism and vaccination, another new development at the time. It is not clear what the editors thought the two subjects had in common, but they must have soon seen that it wasn't much, for only two issues were published.[29]

In a somewhat more positive light, this new world of animal electricity was also the starting point for a long line of science fiction. This still-popular genre got off to a good start with Mary Wollstonecraft Shelley's *Frankenstein* (1816–1818), about a monster brought to life via electricity. Lord Byron, a friend who took part in the initial discussions leading to the story, is said on the other hand to have been more interested in the application of "galvanism" to virility, a promise exploited by unscrupulous quacks.[30]

Later in the century, the mystery and the promise of "galvanism" was still in the air. In a eulogy on Volta, published in 1875, Arago wrote:

I must decline an invitation made me to treat the subject [of the pile] with regard to its medicinal properties and the power it possesses, it is said, of curing certain affections of the stomach and paralysis, for the lack of sufficiently accurate information. I will add, however, that M. Marianini, of Venice, one of the most celebrated physicists of the century, has recently obtained, in eight cases of severe paralysis, results so completely favorable, by a skillful application of electro-motors, that it would be the grossest negligence on the part of the medical faculty not to give their attention to this means of alleviating human suffering.[31]

Medical researchers are now applying electricity to a vast variety of maladies—such as depression, epilepsy, bone repair, paralysis, Parkinson's, and sleep apnea—and it remains an important field of research. We know a lot more today, and also know how much more remains to be discovered and understood.

You're Both Right

Summarizing, then, Galvani had shown in his third experiment that he had discovered animal electricity, and Volta had shown with his pile that Galvani had not done so. How can this be?

In an old Jewish story, told in many ways, an unhappy married couple goes to their rabbi for help. He listens to the husband's side, and says, "You're right." Then he hears the wife's side, and he again says, "You're right." The rabbi's wife objects, "If he's right, then she can't be right, too." The rabbi wrinkles his brow, and finally says, "And you're also right!"

This, in bare essentials, is what happened in the feud between the Galvanists and the Voltaists. For both men *were* right. But both were also wrong. The confusion arose because the true cause of the contractions was not even suspected.

Although Volta had managed to convince the bulk of the scientific world that his instrument confuted Galvani's idea, the truth is that it did nothing of the kind. We know today that electrical activity lies at the heart of all of animal life. Yet after Volta announced his pile, there was virtually no mention of animal electricity in scientific circles for almost two decades.

Then, in 1818, the same year Mary Shelley's *Frankenstein* was published, the Italian physicist Leopoldo Nobili performed a couple of wonderful demonstrations that brought Galvani's idea back to life. That is, he showed clearly the validity of "animal electricity." [32] In one of his experiments, he made a voltaic pile out of frogs. Placing the trunk of one frog on the legs of another, and then building a "pile" in this way, he showed that electric current was produced and, further, that it increased along with the number of elements in the pile. The current was small, but detectable. [33]

There is lovely irony in the fact that the weak current the frog pile generated was detected with a device called a galvanometer. In a further irony, Nobili also found that a current too small to move the galvanometer's indicator would nevertheless cause contractions in a frog.

Though Volta had stood at the doorstep of electrochemistry, he couldn't or wouldn't step through. Though only 55 years old in 1800, he left such discoveries as the electrolytic dissociation of alkaline salts to Davy and others. His interests lay more in beating down the lingering doctrine of animal electricity. Early on, for example, when demonstrating his electric pile before Napoléon, he had tried to give it the appearance of an electric eel by covering it with a skin.[34]

It's probably because of this negative mind-set that he missed out on seeing what was actually going on in Galvani's work. For Volta had seen in his experiments that when a frog was *intact* it could be made to convulse only when hit by a discharge from a Leyden jar or when made part of a bimetallic circuit.[35]

He could have gone further and discovered, as Nobili did two decades later, what has come to be called the injury current. In other words, the animal electricity seen by Galvani in his third experiment was real, but everyone, including Galvani, misunderstood its cause. The current was the result of a bioelectric potential *caused by the injury to the frog's tissues* when it was prepared for the experiments.

Whoever Said Life Is Fair?

Volta, deservedly, was honored, feted, and richly rewarded in his later years. For the last two decades of his life, he had the income of a wealthy man. Upon his death in 1827 a beautiful monument was erected to his memory on the shore of Lake Como, near his place of birth.

It's true that Galvani's name, today, is honored in such terms as galvanometer, galvanization, galvanized iron, and even the nonscientific expression "galvanized into action." But, thanks to an intense feud, plus some curveballs from fate, he had died, in 1798, a bitter, poverty-stricken, and broken man.

There is, however, a certain poetic justice at work. Both Galvani and Volta handed on work to their successors. The electrochemical aspects opened up by Volta's pile have indeed revolutionized the world of science, but have been well mined. The field of electrophysiology inspired by Galvani, however, and its myriad applications to human health, remains an exciting field that is probably still at its very beginnings.

Have we learned anything from the Galvani-Volta feud? How about, "Don't be first"?

Of course, if no one is first, there can be no second.

CHAPTER 3

Semmelweis versus the Viennese Medical Establishment

Childbed Fever

Pathology is all the rage in 19th-century Vienna and every medical student is expected to dissect and carefully inspect diseased organs—which means, often, groping them with bare hands.

An eager young medical student is just finishing a messy postmortem examination. He wipes his hands, wet with blood, pus, and any other manner of secretion—on what? His pants, an apron, his jacket?—then heads off to his next training stop across the street, where he will do an internal examination of a woman who has just delivered a child. His objective is to study the changes that take place in the uterus before and after the patient gives birth.

On the way, maybe he stops to rinse his hands with cold water, maybe not. Perhaps he even washes with soap and warm water. But there's a certain smell on his hands that just doesn't seem to disappear. He wears the smell proudly, for it shows clearly that he is a student of medicine and not of that second-class discipline, obstetrics. The stained clothing, too, has become a badge of honor, among faculty as well as students.

The Allgemeines Krankenhaus

For some medical researchers who broke convention, acceptance and respect came rapidly and easily. Others, like the Hungarian physician Ignaz Philipp Semmelweis, were not so lucky.

Semmelweis began his medical training in Vienna, but feeling like an outsider, and being treated like one, he returned to his native region (now Budapest) for further training. But the primitive conditions of his hometown University of Pest drove him back to Vienna.

Again plagued by snide remarks about his speech and dress, he somehow got through his studies. His native talent and intelligence must have shone through the rough exterior, and by 1846, at the age of 28, he had earned a position as assistant to the director of the First Obstetrical Clinic at Vienna's Allgemeines Krankenhaus (General Hospital). This was a major teaching hospital in the most forward-looking city in Europe, and it featured the largest lying-in clinic in the world.

All was not well, however; Semmelweis found himself looking at a puzzling, and ominous, statistic: an extremely high death rate among the women giving birth in this clinic—averaging some 13 percent, but as high as 30 percent at times.[1]

The Allgemeines Krankenhaus was by no means unique in this respect. In other hospitals and other times the situation was even worse. For three years starting in 1773 disease decimated the lying-in hospitals of Europe, culminating in Lombardy, a region in northern Italy, where it is reported that for one whole year not one woman lived after giving birth in such a hospital.[2]

The diagnosis in almost all cases was puerperal (also called childbed) fever, a ferocious infection that blazed up in the newly delivered mother's body and raged through it like a wildfire, killing within days or even hours. Invariably the patient experienced high fever and pain, but also a variety of baffling symptoms, which might include massive skin eruptions filled with pus and fluid, peritonitis, pleuritis, phlebitis, meningitis, and other "itises" as well, all of which refer to inflammation of the body part named in the prefix. Just defining the disease was a problem.

Puzzled physicians sought causes in every corner. Was overcrowding somehow involved? Some suggested the cause might be "miasma," a kind of noxious air, possibly related to weather changes. Could it be an epidemic type of disease, perhaps related to other major killers like smallpox or the plague? Others blamed all sorts of earthly, atmospheric, or cosmic "emanations"—possibly electrical in nature.

Or perhaps it was the strain to which the young women were subjected by being examined by male students. This explanation was particularly ironic. These women only entered the clinic because, as poor single mothers or perhaps prostitutes, they had nowhere else to go. Though clearly among the socially despised, they were suddenly invested with a class of modesty that even the aristocracy did not claim.

Still, the administration decided to exclude all foreign students from this part of the program, explaining that they probably showed less delicacy than the native Viennese students in their examinations of these women.

But it was all an intellectual exercise, and what appeared to be a natural course of events hardened the hearts of most physicians, who simply felt that this was the way it had to be. Imagine an era of which one professor could write: "The physician should be judged by the extent of his knowledge and not by the number of his cures. It is the investigator, not the healer, that is to be appreciated in the physician."[3]

A Doctor with Both Heart and Brain

Semmelweis was different. He heard the cries of the destitute, desperately afraid women. Several times a day a priest in full canonical dress, preceded by a choirboy ringing a small bell, passed through the corridors to give the last sacrament to the dying. The priest's bell became for Semmelweis "an exhortation to search with all his energy" for the real cause of this matricidal slaughter.[4]

Curiously, in the Second Obstetrical Clinic, a different lying-in section of the same hospital, the mortality rate was far lower, on the order of 2 percent. The First was the teaching section for the medical students, while the Second was used for the instruction of midwives. The difference in mortality was recognized, and Semmelweis saw young, pregnant women who had mistakenly applied for admission to the First Clinic down on their knees begging for transfer to the Second Clinic.

Semmelweis read and listened to all the guesses regarding cause. But nothing seemed to make sense to him. He did learn that although the hospital opened in 1794, medical students did not do their own postmortem dissections until 1822. It was not long afterward that the ominous, puzzling rise in the death rate began. And not until 1840 were the medical students separated: male students were assigned to the general hospital (First Division), while the females, invariably midwife students, were assigned to the lying-in section (Second Division). But the male medical students also delivered babies in their own division.

Then, in March 1847, Semmelweis went off on a vacation. On returning he found that Death, ever close by, had struck down his close friend and colleague Jacob Kolletschka, a professor of forensic medicine at the hospital. Kolletschka had sickened and quickly died after his finger was accidentally punctured with a knife during a postmortem examination.

The autopsy revealed important information to Semmelweis. He realized that the changes in Kolletschka's body were the same as those he had seen in the bodies of the women dying of puerperal fever.

Day and night the vision of his dear friend's corpse haunted him. But now Semmelweis's earlier training in pathology came in handy. He had studied with the highly respected Bohemian pathologist Karl von Rokitansky. This eminent professor of pathology had maintained that symptoms of disease are the external manifestations of disease in organs and tissues and not, say, a punishment by God because the mother had not carried out her religious duties faithfully.

Cleanliness

Semmelweis reasoned that such causal explanations as overcrowding, modesty, and miasma were all equally unreasonable, that these conditions held just as well at the Second Clinic. He began to suspect that cleanliness, or lack of it, was somehow involved.

Note that he was by no means the first to come up with this idea. Other times and other places had seen an emphasis on cleanliness: The biblical Jews speak often of hand washing and cleanliness in their writings and teachings. It shows up in very early Indian literature; in Babylon and Egypt; in 10th-century Persia; and in some medieval European medical literature as well.

From 1742 to 1758, Sir John Pringle, physician-general of the British army, did studies on antisepsis, infection, putrefaction, fever, and contagion, which he communicated to the Royal Society. In 1752 he strongly advocated the "gospel of cleanliness" in his *Observations on the Diseases of the Army.* Medical historian Roy Porter writes of the work: "While not strikingly original, it captured the Enlightenment concern for hygiene, public health and the value of life."[5]

But most surprisingly, in 1843, American Oliver Wendell Holmes, jurist, physician, and writer, even suggested infection as a possible cause of childbed fever. A powerful writer, he described the infection as being carried by attendants "from bed to bed as rat-killers carry their poison from one household to another."[6] But Holmes was better known as a writer and university professor at Harvard's Medical School than as a practicing doctor, and his generalized warnings about cleanliness sank with little effect. The English argued that they were already careful enough in this respect.

At the same time, Europe's cities were growing rapidly. As cities grew, so did the number of the desperately poor. And, as Charles Dickens has described so graphically, they had little in the way of a safety net. Charity and state hospitals became the last resort not only of the sick but also of the poor parturient women who could not afford the services of private physicians and experienced midwives.

Though hardly a desirable haven, such hospitals were, these desperate women felt, better than the alternatives—an even more dangerous abortion, trying to find a taker for the child, or, too often, murder of the infant. If the woman gave birth in a state hospital, then the state would accept the child into its care even if the mother was able to leave the hospital under her own power.

But as the dirt of the poor moved into the hospitals along with the poor themselves, it must have seemed rather pointless to worry too much about cleanliness.

Semmelweis began to suspect the hands of the students and the faculty. These, he realized, might go from the innards of a pustulant corpse almost directly into a woman's uterus. He theorized that the basic cause of puerperal fever was some sort of septic infection—from "poisoned" cadaveric material absorbed into the victim's system through the vascular system. In Kolletschka's case the entrance to the bloodstream was the knife wound; in the women, it was the "lacerated" uterus—that is, the raw surface left in the mother's uterus after her child is delivered.

Even when the practitioner's hands were washed, Semmelweis guessed, this was not enough to remove "invisible cadaver particles" that clung tightly to the skin. He decided to test the idea, but, not knowing what he was dealing with, he could only carry his idea to extreme lengths, and he became obsessed with cleanliness. Starting in May 1847 he put through a rule calling for, first, washing the hands with soap and water, including a scrubbing with a nailbrush, then following up with thorough washing with a chlorine solution—until the hands became slippery and the cadaver smell could no longer be detected. This was especially important, he felt, before any contact with the next patient, especially if it was a woman about to undergo, or one who had just undergone, childbirth. (In the Second Clinic the practitioners were not dealing on a regular basis with all manner of infectious diseases. So even if the midwives didn't wash their hands on a regular basis the damage was not as severe.)

Within a month the mortality rate in the First Clinic dropped to 2 percent. Three eminent physicians—colleagues of Semmelweis at the hospital—quickly accepted his thesis. He was on his way to glory.

Guilt and Resistance

But now a terrible irony began to assert itself. Semmelweis himself had done many postmortem exams. Acceptance of his idea carried with it the implied fact that the longer the medical practitioner had been doing his thing, the more deaths sat immediately on his shoulders. "Consequently," he wrote to a colleague, "must I here make my confession that God only knows the number of women whom I have consigned prematurely to the grave. I have occupied myself with the cadaver to an extent reached by few obstetricians."[7] Semmelweis accepted this but felt a heavy guilt, a guilt that may have played a part in his later decline.

Few others in the haughty Viennese medical establishment were willing to accept that terrible burden. Also, who was this young whippersnapper—a Hungarian to boot—to tell them they were killing people?

And, in truth, was their negative reaction so strange? After thinking in terms of deep, even mystical possibilities, ranging from cosmic influences to strange emanations, what could they think of the idea that washing one's hands was going to cure puerperal fever? What, after all, did medicine know of cures? There were no antibiotics; sterilization, even the realization that germs cause disease, still lay in the future.

Some of those involved, both students and faculty, were angered at being made to look ridiculous, others figured Semmelweis had simply gone over the hill. Even patients found a negative angle. A rumor started in the clinic that his treatment was really an insult, implying that the students were washing their hands because the women in this ward were especially dirty.[8]

A Serious Error

Still, the results were clear. Unfortunately, Semmelweis now made the first of several serious errors. Though he had in his hand strong evidence for the efficacy of his method, he chose not to communicate the results to the official medical establishment of Vienna.

His reasons for hesitation are perhaps understandable. Born and brought up in Hungary, he never felt comfortable, either in speech or writing, with the Viennese German that was the language of his peers.

Some of this insecurity probably had to do with personal factors. The son of a grocer, and from a "back-woods region," he was not exactly the autocratic Viennese's idea of a peer. As far as they were concerned, the only thing lower than a Magyar (Hungarian) was a Magyar Jew. Semmelweis, a Roman Catholic, may have chosen to keep a low profile out of some fear that his Germanic-sounding name could even lead to his being taken for a Jew.[9]

His fear was probably not imagined. Medical historian Victor Robinson wrote in 1912: "Even so well-liked a teacher as Professor Nothnagel lost his popularity when he tried to combat the anti-Semitism of the students; the classes rioted, and for a time all the courses in medicine were suspended."[10]

But, as always, there were exceptions. Ferdinand von Hebra, a well-established colleague and strong supporter of Semmelweis's work, was Jewish, which didn't seem to hurt him professionally.

There is, in fact, an even greater irony here. Vienna was indeed at the forefront of medicine; and the establishment was indeed in the hands of native Viennese. But, ironically, the men who brought glory to Viennese medicine were almost invariably not Viennese; they were mainly Slavs, with a sprinkling of Magyars. Of those I mention in this chapter, Rokitansky was a Czech, Skoda a Bohemian, and Kolletschka and Hebra were both Czech Jews.[11] In some crazy way, this latter fact probably strengthened rather than weakened the anti-Semitism.

Finally, Semmelweis's chosen field, obstetrics, was also a problem. He had been turned down earlier for study in other fields, in pathology and then in medicine, and thereby ended up in obstetrics. But obstetrics was at the time considered a lowly occupation for a doctor, more fit for the midwives who did most of the deliveries.

This is not to say that his treatment regimen had no publicity at all. Hebra, who was the editor of a prominent Viennese medical journal, published brief notices of Semmelweis's results in the journal in December 1847 and April 1848. At least two other of Semmelweis's followers published lectures on the subject.

In fact, one scholar, Erna Lesky, maintains: "Scarcely ever had a discovery attained such quick publicity as did that of Semmelweis."[12] But it was nowhere near enough, and the movement stagnated.

Semmelweis and his students tried sending out detailed letters to several prominent medical practitioners, inviting response. One of these letters found its way to Professor Gustav Adolph Michaelis of Kiel, Germany, and had a terrible outcome. A highly respected and conscientious teacher of obstetrics, Michaelis, like Semmelweis, was extremely perturbed by the toll taken by puerperal fever. In fact, unable to cope with the carnage, he had earlier chosen to close his hospital temporarily. He introduced Semmelweis's method and saw immediate improvement.

But, at the same time, his own niece had already succumbed to the scourge. Unfortunately, Michaelis had delivered her and, now seeing that the disease was preventable, took on the same guilt that Semmelweis fought. Michaelis threw himself under a train, leaving only his learned book, *The Contracted Pelvis,* to keep his name alive.[13]

The publicity achieved by Semmelweis's few remaining supporters had little effect. How could this be? The question becomes even more puzzling when we compare Semmelweis's situation with those of Harvey and Galvani. Delayed acceptance in the latter two cases was of little medical consequence: there were no immediate medical casualties. But in Semmelweis's case vast numbers of women were dying unnecessarily, and there was a method that had been shown to work. We can only guess at the enormous frustration he must have felt.

But now, as in Galvani's case, factors other than medicine and reason arose and took over.

Fate Takes a Hand

The year 1848 is marked indelibly in every textbook of European history. A series of rebellions marked a widespread revolt of the new against the old.

In Vienna—capital of the Austrian Empire—and elsewhere there were the beginnings of liberalization. Angry mobs forced the hated Prince Metternich to abdicate and flee to England. Treaties and undercover agreements abounded; practically all of Europe was in an uproar. Capitalists, monarchists, democrats, and religious leaders all jockeyed for power. The reform movement, which Semmelweis and his colleagues joined, erupted in the academic and medical worlds as well. Sides formed in large institutions like universities and hospitals and attempts were made to root out the old reactionary leaders.

But across Europe the established forces dug in and the power structure held; with only a few exceptions, the reform movement went down in flames.

Many of the extreme revolutionaries in Vienna fled to England and the United States. The power structure at the Allgemeines Krankenhaus emerged unscathed. Semmelweis was permitted to return to the hospital clinic. But his superior, the reactionary Professor Johann Klein, saw his assistant in the clear robes of the revolutionaries.

Astonishingly, Semmelweis noted, during the month of March 1848, when the excitement there was at its peak, and little teaching or research had been done at the hospital, not a single woman had died in childbirth in the First Clinic.

Resistance

A year later, positive reaction to Semmelweis's thesis seemed to be building. In March 1849 his temporary position as assistant to Klein came up for renewal. He applied for another two years. In spite of, or perhaps because of, the initial positive reaction to Semmelweis's ideas, Klein, with the support of the "old guard," chose not to renew his appointment. They had not forgotten that Semmelweis had played a part in the revolutionary Academic Legion. Appeals by Rokitansky, Hebra, and Skoda were of no avail.

As an alternative he petitioned for a position as *Privatdozent* of midwifery; he would be an independent physician with hospital privileges and permission to teach students privately. His friends at the hospital assured him he was a shoo-in. But month after month the appointment never came. Again, appeals by his colleagues were unavailing. He supported himself with some private patients; to help pass the time he performed some animal experiments as a way of building up supporting evidence for his cause.

Support came in other ways. Dr. Karl Haller, an eminent and influential physician, invited Semmelweis to lecture on his work to the Vienna Medical Society. He hesitated, but finally appeared in May 1850. He did well; it was a great triumph for the humble Hungarian. But then he lost another good opportunity. Asked to prepare a paper for publication, he declined, apparently feeling even more insecure with his written German than with his abilities in speaking it. Because he did not follow through with a written presentation, his side of the lecture and ensuing debate appeared only as a brief summary, whereas his opponent's presentation got a full-scale treatment.

On top of this, his lecturing at the society was enough to further irritate his still powerful opponents, who pulled out all the stops—social, administrative, medical—of which they had control. They cut him socially and, more important, professionally. When his application for *Privatdozent* was finally approved in October 1850—17 months after he applied—it had some strange strings attached. Bad enough that it was unpaid, it also included some galling restrictions. He could demonstrate only on a mannequin, though there were plenty of corpses available, and he could not grant certificates of attendance, a common part of the job.

Was this a deliberate attempt to upset him? Probably. In any case, it worked, and it resulted in another terrible mistake on his part.

Flight

Angry, distraught, suspecting even his friends of deserting him, he simply packed up a couple of days later and left for home. Feeling perhaps that his three major supporters—Rokitansky, Skoda, and Hebra—were somehow involved in the insult, he never even said good-bye to them. They in turn felt he was leaving them in the lurch.

Homecoming, after 12 years away, presented its own difficulties. After an initial period of recognition in Vienna, here he was crawling back to a medical backwater (at least compared to Vienna).

Along with other sufferers under Austrian rule, Hungary had made an attempt to free itself in 1848 and 1849, but the rebellion was stamped out there as well and its citizens lost any constitutional rights they once had. The leaders of the revolt, statesmen and officers alike, were being court-martialed and shot; the flower of Hungarian youth, many of whom had also participated, were being enrolled in the Austrian military and dispersed to outlying areas of the still-extensive empire.[14]

So while Vienna actually saw some academic freedom as a result of the 1848 activities, the situation at the University of Pest had deteriorated. As a result, the academic standards were even worse than Semmelweis had remembered.

His parents were dead; several brothers, who had taken part in the revolution, were refugees. The only family he could turn to were a sister and one brother, a parson.

After a while, things took a turn for the better. In May 1851 he obtained a post (again, unpaid) in the Obstetric Division at St. Rochas Hospital in Pest, and in short order he instituted his method of chlorine disinfection. Results were as gratifying as they were in Vienna, and over the next few years his situation, including his private practice, began to pick up.

In July 1855, Semmelweis was appointed professor of theoretical and practical midwifery at the University of Pest. His fame in his home region grew satisfactorily, and so did his ego. Things were looking good. He met and married Maria Weidenhofer, who was to provide him some needed support and the small amount of peace he found in his tortured life.

He even began speaking and writing about his ideas. His first real publication, "The Etiology of Childbed Fevers," appeared in a Hungarian journal in 1858.

Still, his doctrine was making little headway elsewhere. Justus von Liebig, a tempestuous but outstanding chemist, was insisting that infection had chemical, not biological, underpinnings. This led a prominent Vienna medical weekly to urge an end to the chlorine treatment. The mortality rates at the Allgemeines Krankenhaus rose again into the teens.[15]

Rudolf Ludwig Carl Virchow, the eminent German pathologist and founder of cellular pathology, stood up at a major meeting in Berlin and lent his considerable influence to the anti-Semmelweis crusade.[16]

Semmelweis felt he had to do something to help staunch the continuing pandemic of childbed fever. He decided, too, that it was time to put Hungary, and especially Budapest, on the medical map. It was time to put all of his thinking together into a book, and in German.

So, in a fascinating reprise of the first two chapters in this book, our hero hesitates for a dozen years or so, dithering, equivocating, before publishing in full his carefully spelled-out thesis. It was not until 1860–1861 that he generated and finally issued his magnum opus, *The Etiology, the Concept and the Prophylaxis of Childbed Fever.*

The first half is a dense, not very orderly discussion of his regimen: extensive tables on the mortality experiences at the Allgemeines Krankenhaus; details of other epidemics in other institutions; discussion of other theories. A modern biographer, Frank G. Slaughter, describes it as a massive compilation "almost pathetic in its painstaking thoroughness." But, Slaughter adds, "he made no assertion without drawing up a battalion of proofs to sustain it."[17]

In the second half of the book, Semmelweis attempts to counter all significant attacks of which he was aware. But his attempt to deal with all the many arguments, and with the many arguers as well, leads to an avalanche of repetition and convolution.

For example, Carl Braun was the obstetrician who replaced Semmelweis as assistant to Klein at the Allgemeines Krankenhaus, and eventually succeeded Klein himself. Directly contradicting Semmelweis, Braun had earlier published a long paper on puerperal fever and listed some 30 different possible causes of the malady.[18] In the *Etiology*, Semmelweis answers each of Braun's 30 purported etiological factors.[19]

The tenor of some of his remarks, however, froze the blood of all who read them, and continues to raise questions today. He concludes this particular section on Braun with: "Oh Logic! Oh Logic . . . we recommend most urgently that if he [Braun] should again feel called upon to fight for the epidemic deaths of maternity patients, he should first attend at least one semester of logic."[20]

Semmelweis referred to Friedrich Wilhelm Scanzoni, a prominent professor of obstetrics at the University of Prague, as a "wretched observer."[21] He also noted that "823 of my students are now midwives practicing in Hungary. . . . They are more enlightened than the members of the Society of Obstetrics in Berlin; they would laugh Virchow to scorn if he attempted to lecture them on epidemic puerperal fever."[22]

Not surprisingly, among those who counted most, reaction to Semmelweis's *Etiology* was, mainly, icy rejection. Few readers had the interest or energy to plow through the dense combination of theory and statistics in the first half; the book's second half, at least the good parts, undoubtedly got greater readership, but was not likely to inspire medical reviews. His supporters cringed; opponents were enraged and did all they could to suppress the book. Only a few short notices appeared in the medical press, most of them uncomplimentary.

So strong was the negative reaction that in some cases potential supporters feared for their careers. A possible influential supporter, Carl Mayrhofer, for example, felt that he had to keep his beliefs quiet. For he worked under Carl Braun, still one of Semmelweis's most vociferous critics. Eventually Mayrhofer published his views. These were angrily denounced, and his own career began to suffer.[23]

Once again, the emotional Semmelweis's high hopes were dashed. How much can such a man take? Unable to restrain himself and virtually barred from publication in the major medical journals, he lashed out even more violently at several important detractors in a series of open letters. That is, he sent copies of the letters around, and published at least two in a pamphlet.

To Dr. Joseph Spaeth, professor of obstetrics at St. Joseph's Academy, part of the University of Vienna, he wrote: "You, Herr Professor, have participated [in the massacre]. . . . This homicide must stop."[24]

And to the prominent Friedrich Wilhelm Scanzoni, he scolded: "Your teaching . . . is founded on the dead bodies . . . of women murdered through ignorance. . . . I have devoted page 103 of my book to demonstrate the fatal mistakes you made in the question of puerperal fever. . . . Should you . . . continue . . . to teach your students the doctrine of epidemic puerperal fever, I denounce you before God as a murderer. . . ."[25]

Now and then he managed to moderate his tone. To Scanzoni he also wrote: "I entreat you . . . to acquire an intimate knowledge of the truth as it is set forth in my book, so that according to your kindly disposition you may be able to find support for new opinions. . . ."[26] This had no more effect than the other approach.

Objections: Reasonable or Unreasonable?

As we saw earlier, the objections Semmelweis faced were not completely bullheaded and childish. But they do show how hard it is for a new, or at least challenging, idea to find its place in an ongoing show. Even the microscope fell into this category. The first compound (multilens) microscope dated back to the early 1600s, and Leeuwenhoek had already seen single-celled organisms with his device. Yet even so forward-thinking a physician and teacher as Rokitansky at first abjured its use. He neglected to mention it in his influential, multivolume treatise on pathology (1842–1846).

Professor O. H. Wangensteen believes: "Had Rokitansky counseled Semmelweis to examine microscopically the lochia,[27] he probably would have observed the organisms which . . . Pasteur later (1879) labeled as the cause of puerperal fever. Yet [even] Pasteur's announcement did not find ready acceptance among obstetricians in Paris."[28] Ironically, Mayrhofer apparently had identified bacteria in puerperal fever even earlier than that, perhaps as early as the beginning of the 1870s.[29]

Carl Braun did, early on, suspect that some sort of infectious agent lay at the heart of puerperal fever. But he believed it was transferred through the air (the miasma connection again) and he had an improved, and expensive, ventilation system installed at the Allgemeines Krankenhaus maternity clinic in 1864. Later in the same year one of his students presented papers arguing that improved ventilation, not chlorine washings, had helped lower the mortality rate in the clinic.[30]

Reprise

Interestingly, the controversy engendered by Semmelweis's book seems to have erupted again in our own era; once again Semmelweis's *Etiology* is receiving some strongly negative reviews.

Medical historian Sherwin B. Nuland wrote in 1979, "The first part of the book deals in a rambling, often repetitious, fashion with his theory. In the second part . . . Semmelweis delivers himself of a long polemic in which he not only answers all of his major detractors, but violently attacks most of them. His discussion is frequently interrupted by brief torrents of abuse." [31]

Frank P. Murphy, who did the first translation of Semmelweis's book into English in 1941, felt that "if Semmelweis had only spent more time in clearly stating his views and less in argument, his book would be twice as good and half as long!" [32]

Howard W. Haggard, author of the 1929 classic *Devils, Drugs, and Doctors,* wrote, "If Semmelweis had wielded the pen of Oliver Wendell Holmes his great discovery would have been adopted throughout Europe in the first twelve months after it was made. . . ." [33] Gerald Weissman, writing recently in the *Lancet,* seconds the idea. [34]

On the other hand, there are medical thinkers, including some in Semmelweis's own day, who would disagree with these evaluations of the *Etiology.* In fact, the initial response to it, especially in Germany, [35] was actually very promising. One colleague, a Dr. Kugelmann from Hanover, wrote to him: "Permit me further to express the holy joy with which I studied your work. . . . It has been vouchsafed to very few to confer great and permanent benefits upon mankind, and with few exceptions the world has crucified and burned its benefactors. I hope you will not grow weary in the honorable fight which still remains before you." [36]

In Berlin, another contemporary, Dr. Max Boehr, presented a paper before an obstetrics congress and drew heavily on Semmelweis's work. In a specific reference to the *Etiology,* he mentioned that "the superstitions of our predecessors, who believed in unknown cosmic-telluric-atmospheric influences, were dealt a severe blow, as was the belief in miasmata. . . ." [37] Apparently Kugelmann and Boehr managed to understand Semmelweis's prose.

At the beginning of our own century, Victor Robinson, author of a highly successful history of medicine, wrote: "As far as its scientific value is concerned, no praise can be too high: page after page could stand, without revision, in the most modern treatise on the subject." [38]

And Henry E. Sigerist, a highly honored medical historian,[39] describes it as "simple and unadorned, but cogent through its presentation of hard facts and through its inexorable consistency—being a work which was intended to overcome the last resistance to his teaching."[40]

Finally, and most telling, K. Codell Carter, a Semmelweis specialist at Brigham Young University who has done much work in this area, including a recent translation of Semmelweis's book, feels it "is surely among the most moving, persuasive, and revolutionary works in the history of science."[41]

Clearly not everyone felt, or feels, the *Etiology* was unreadable. Difficult, perhaps. But not unreadable.

Bad Timing

Returning now to Semmelweis's day, what was the real problem? Is it possible to discern a "real" cause underlying the resistance to his idea?

Brian Inglis, another medical historian, maintains: "He happened to produce his theory at the worst possible time—when professional complacency was still entrenched, before Pasteur's work was known. And Semmelweis *was not prepared to obey the conventions of polite medical disputation . . .*" (my italics).[42]

So, was that a good reason to ignore or combat a doctrine that had been demonstrated to save women's lives?

Toward the end of his life, Semmelweis suffered a severe mental breakdown. Was it an obvious result of the reception this highly emotional personality received at the hands of his peers? Was it a response to the constant struggle and rejection? It seems very likely. Some reports say he began acting strangely, even wildly, sometime around the middle of 1865.

Yet even here there are alternate views. Nuland writes: "By 1862, when he wrote the last open letter, Semmelweis is described as having bouts of depression alternating with periods of elation. Although he could still carry out his professional duties, he was suffering from problems with his memory and fits of bizarre behavior."[43]

Citing work by a Yale neuropsychiatrist, Dr. Elias E. Manuelidis, he suggests that "the memory loss, hyperactivity, and marked change in physical appearance that were prominent in Semmelweis's case are so characteristic of Alzheimer's disease that they argue convincingly in favor of that diagnosis. The pathological changes in Semmelweis's brain also support this contention of presenile dementia. . . ."[44]

The difficulties in teasing out the truth can be seen in Nuland's evaluation of the opposing, more sympathetic, point of view: "Rendering the true sense of Semmelweis's labored German into effective English," he writes, "has proved inordinately difficult." Of the two Hungarian authors, Gortvay and Zoltan, who did what he considers the best translation, he adds, "they maintain that Semmelweis was entirely sane until the last few weeks of his life, in spite of abundant evidence to the contrary. . . . [But] their aim as Hungarian academicians is to interpret the events of Semmelweis's life in the most salutary way possible." [45]

Nuland and Weissman seem also to be saying that Semmelweis's problems were brought on by his own attitude. Nuland feels that Semmelweis's tragic fate was more like that seen in the dramas of Sophocles (self-caused) than Aeschylus (cosmic destiny). [46] I consider this evaluation far more drastic and harmful than their medical observations.

Are such revisionists perhaps worried that the adulation for Semmelweis had gone too far? In our own century, his name began to take on a rosy glow. In 1949, Morton Thompson wrote a best-seller, *The Cry and the Covenant,* which sold over a million copies. It was also praised in glowing terms: Edmund Fuller, in the *Saturday Review,* called it a "noble book, deeply conceived and powerfully written." Paul de Kruif wrote in the *New York Times,* "This book is an event in literary history . . . the first novel truly to describe medical discovery and to tell truly the martyrdom of a discoverer. . . ." [47]

Frank G. Slaughter's highly adulatory biography of Semmelweis, which appeared only a year after Thompson's book, also did very well.

I would argue that putting the blame on Semmelweis for his inability to get the message out distorts the picture of what actually happened. Suppose his treatment had gathered enough momentum to be widely accepted much earlier in his career. It seems unlikely that he would have lashed out as he did. The fault, if there is one, lies in the complexities introduced by any new, and especially revolutionary, idea.

In any case, it was not until the later work of Koch and Pasteur, which showed conclusively that infections, including childbed fever, were caused by germs, that Semmelweis's star could rise and hang high. Even some of his most bitter detractors began to come around. But for him it was already too late, he died on August 13, 1865, before any of this took place. Like Galvani, he died a defeated and bitter man.

Death

Semmelweis concluded his *Etiology* with: "If I am not allowed to see this fortunate time [the conquest of puerperal fever] with my own eyes . . . my death will nevertheless be brightened by the conviction that sooner or later this time will inevitably arrive."[48]

It may be that he didn't even have this comfort. Two stories about his demise have come down to us, both unhappy. Both agree that his mental condition deteriorated rapidly, starting, his wife later said, on July 13, 1865, and that he was committed to a mental hospital on July 29, 1865,[49] at the age of 47.

After that the picture becomes less clear. One common depiction suggests that he died of the same sort of bacterial infection that is associated with puerperal fever, possibly caused by an injury to one of his fingers from an examination he made when not yet committed.

This interpretation is found in many sources, particularly the older ones. But unexplained factors in his death have led to two new developments. First, Semmelweis's remains were dug up in 1963 and reexamined. Then, in 1977, Georg Silló-Seidl, a Hungarian physician and writer, dug deep and came up with a collection of documents concerning Semmelweis's final illness and death. This has led to a new, and far darker, interpretation of Semmelweis's last few weeks of life. The following information is drawn directly from a 1995 paper on the outcome of the new inquiries.[50]

First, it seems that the Viennese asylum to which Semmelweis was brought was by no means the best in that city, and that he was tricked into going there by being told that he would be visiting the working quarters of his old friend Ferdinand von Hebra.

Second, of the three physicians who signed the commitment paper (required by law for involuntary commitment), not one had been trained in psychiatry. Also, even though there were respected psychiatrists in Budapest, there is no sign that any were consulted.

Third, there is nothing to indicate that a priest was called in to administer the final sacraments.

Finally, and worst of all, the paper's authors state:

The autopsy revealed major injuries that could only have been sustained in beatings to which Semmelweis had been subjected while in the asylum. There were serious injuries involving even the bones. . . . The cause of death was identified as pyemia [a massive blood infection]. Given the autopsy report and the medical record of Semmelweis's stay in the asylum, it seems most likely that Semmelweis was severely beaten by asylum guards and then left essentially untreated.[51]

His funeral was attended by, among a small handful of others, Karl von Rokitansky, his early teacher and mentor; the same Joseph Spaeth to whom he had written one of his open letters, and who was just coming around to Semmelweis's side; and Carl Braun (still one of his bitterest enemies). It's hard to understand Braun's motive for attending—particularly since, according to K. Codell Carter, "the intensity of Braun's feelings may also be revealed in the fact that between the years 1855 and 1881, he did not mention Semmelweis by name, not even when discussing views he had earlier ascribed to Semmelweis."[52]

From Budapest, only Lajos Markusovszky, a longtime friend and supporter of Semmelweis, attended. Not a single member of Semmelweis's family was in attendance. His wife later explained that her absence was due to illness.

Even in his own city, the local newspapers virtually ignored him, and when two of his assistants applied for the newly vacant job at St. Rochas, neither one got it. The man who did get the job had no training in obstetrics, and not long after he took over, the mortality rate shot up to more than six times the rate Semmelweis had managed to achieve.

By 1891, Hungarian authorities had become aware of a monumental mistake. They removed his body from an obscure Viennese cemetery, over the objections of the Viennese, and reburied him in Budapest. The Viennese and the Germans at this point were calling the "Pest Fool" an honorary German. In 1906, Hungary honored him with a statue located in a small square in central Budapest. It shows him standing tall, his book under his arm; on the pedestal sits a woman, infant in her arms, gazing at her benefactor.

After Semmelweis's remains were examined in 1963 they were, finally, reinterred in the courtyard of the house in which he was born, which is now a medical history museum and is named in his honor.

Semmelweis was not the first person in the medical field to be bitten by the cleanliness bug. But he made the most noise, and he suffered for it. Dr. Constance E. Putnam, of the Wellcome Trust in London, would turn the noise idea around. She suggests that part of the problem was that he did not make enough noise—that is, he did not publish his work as early or as carefully as he should have.[53]

Either way, Semmelweis may deserve credit for more than the cleanliness idea. K. Codell Carter points out that by 1850, Semmelweis had broadened his net to include, as causally significant, every kind of putrid matter, meaning from living diseased organisms as well as from cadavers. This made it possible for him to explain every case of childbed fever.[54]

But it also created a connection with another widespread problem. Called surgical fever at the time, this disastrous result of wounds and cuts is what we would now call blood poisoning. Thanks to Semmelweis, both puerperal and surgical fevers came to be recognized as infected-wound diseases. And, Carter argues, this understanding paved the way for what may reasonably be called the most significant development in medicine: germ theory, the idea that disease is caused by miniscule living organisms.[55]

Germ theory also cleared the way for an acceptance of Semmelweis's ideas, since it finally provided an understanding of his "cadaveric particles."

And so we have this obscure Hungarian obstetrician, laughed at and scorned by Braun, Virchow, Spaeth, Scanzoni, and others, who not only found a way to save millions of women, but who also helped lay the foundations for what was perhaps the most important single development in all of modern medicine.[56]

Possibly, as Nuland and many others maintain, Semmelweis brought his misfortunes on himself. But decorum and timidity didn't seem to be working very well, either. Oliver Wendell Holmes, after all, had a similar idea even earlier than Semmelweis. He also made some good suggestions as to how to deal with the problem. But after sending up his balloon, he moved on to other things—and lived a long, rich life.

More likely, I think, is that if Semmelweis had not played the bulldog, had not responded with so much noise (including the noninformational kind), and had not attacked and flailed with wildness and venom—thereby engaging some who might have just ignored the whole situation—the field might have taken even longer to progress than it did, taking countless lives that have instead been saved.

CHAPTER 4

Bernard versus Chemists, Physicians, and Antivivisectionists

Experimental Medicine

Claude Bernard, a 19th-century French physiologist, performed one of the great tricks of all time: the application of science to the art of healing. He showed that medicine based on experiment and fact was superior to that based on theories and historical precedent. He showed, too, that in order to achieve that goal physicians needed a far deeper knowledge of what goes on in the body—in the healthy body as well as the sick one. And he created, virtually single-handedly, a method for doing all this.

Shortly after his death in February 1878, the *Popular Science Monthly,* a British journal, published a sketch of his life which stated: "His name is connected with important discoveries in nearly every department of human physiology, and the influence of his method has borne fruit wherever this science is studied." [1]

In the latter years of his life, he was showered with honors, was revered by those colleagues who knew him personally, and was a member of most of the learned societies of the Western world. Compared with the experiences of Galvani and Semmelweis it would seem that, by comparison, Bernard had it all.

Another View

But, writing in the same journal just three months after the first memorial sketch appeared, Karl Vogt (1817–1895) saw things very differently. A physician who had known Bernard in his early days in Paris, Vogt noted that he had, after several decades, run into Bernard again, when Bernard was in his 60s. At this point, Vogt wrote, he would have expected Bernard to "feel happy, seeing that everything which the ambition of the *savant* could wish for was offered to him, the highest positions at the universities and in the learned societies, a seat in the [French] Senate, for which he was indebted to his scientific eminence, and not to any political services. . . ."

And yet, Vogt added,

> he was weighed down, on the other hand, by serious bodily ailments and by the saddest of domestic misfortunes. At first I did not recognize him when, pleasantly and kindly, as of old, but gray-haired, and with his head inclined on one side, he stepped up to me at a provincial meeting, and reminded me of the old times in the Rue Copeau and the Pitié Hospital. "I have passed through a great deal since that time," he said to me, "which may have left some traces in my appearance—for I notice that you look at me in surprise—but let us chat about those times and about our old friends, it does me good."[2]

Vogt had it right. Bernard had paid dearly for the many honors showered on him. A relatively small part of that payment was in money problems; more important were problems with his health that surfaced in the spring of 1860, at the young age of 46; but hardest to take, and perhaps basic to the other two, were the unrelenting disparagement he was forced to endure strictly because of the route he had chosen for his scientific research, responses that affected his work, his career, his health, the memory of his name, and even his domestic life.

Understanding what happened to Bernard and how he overcame opposition that would have quickly floored a lesser man requires a brief look at his formative years.

Early Years

Bernard was born on July 12, 1813, in the village of Saint-Julien, in the winegrowing region of Beaujolais. By his teens, his family could no longer pay for his schooling and he was apprenticed to a pharmacist. Considering how much Bernard was able to accomplish in his later years, we can only speculate what kind of agony he must have suffered in the two years he spent at this trade. Consider the following story, told by Sir Michael Foster, one of his biographers, about Bernard's employer.

As was usual at that epoch the clients of the shop, especially the old women of the outlying villages, made a constant demand for a syrup which seemed to cure everything, and Bernard, to his astonishment, found that this favorite syrup was compounded of all the spoilt drugs and remnants of the shop. Whenever Bernard reported a bottle of stuff had gone wrong, "Keep it for syrup," replied the master; "that will do for making syrup."[3]

We do know that Bernard was turned off by his experiences there, and that his young yearnings lay elsewhere, specifically in the world of writing. He tried his hand with a light play. Receiving some encouragement, he went to Paris with hopes of having another play, this one a tragedy, performed there.

He managed to show it to one of the major critics of his day, Saint-Marc Girardin. Girardin saw some merit in it, but not enough to encourage Bernard in his artistic ambitions. Learning of his earlier experience in pharmacy he suggested that Bernard study medicine—a pregnant suggestion if ever there was one. Perhaps Girardin's name should be honored right alongside Bernard's.

By 1839, Bernard was interning at the Paris municipal hospitals, where his talents began to emerge. His writing talent, too, was quickly put to use when he published a large and beautifully illustrated volume on operative surgery in collaboration with a colleague, M. Huette.[4]

But, very quickly, his real interests turned to anatomy and physiology. Anatomy had long been a science and was both well advanced and well presented. But physiology—that is, the *workings* of the body—was a mass of inexact and uncorrelated data.

A major reason for this was the biomedical world's belief that structure and function in the human body were perfect correlates. The general idea was that you could deduce the functioning of an organ from its anatomy, so why bother with experiments on organ functions? One French researcher, E. J. Georget, even argued that nutrition was not a function because it lacks a specific organ.[5]

Entire systems of function were built up in this way. Oxygen, which had been discovered a century earlier, was known to be a factor in combustion. Because it is drawn into the lungs, that must be where sugar is burned to produce the body's heat and energy. It took years for Bernard, who initially began experiments to back up this idea, to beat it down.

Though Harvey had opened a door to use of experimental methods in physiological inquiry two centuries earlier, vitalism remained a force in explanations of life's operations. Anything that was not understood was put down as an example of a "vital force," which meant it didn't have to be explained. As was the case with Harvey, vitalistic theory still stood in the way of sensible investigations at almost every turn.

Inspiration

As Bernard entered this world, he was struck by the work of François Magendie, who was fighting a rearguard action against these ideas. A physician and surgeon at Hôtel-Dieu Hospital and professor of medicine at the Collège de France, Magendie was a modified vitalist; although he accepted the existence of vital forces, he insisted that some areas of living activity could be opened to experimental study. He began the first journal of physiology in 1821, and demanded that all "facts," even those that were widely accepted, should be subjected to experimental verification.

From 1841 through 1844, Bernard carried out the duties of *préparateur,* assisting Magendie in experiments, often including vivisection activities, at the medical school. This early work convinced Bernard that vivisection (controlled experimentation on living animals), was the best route to a solid understanding of life's processes, possibly even the only route.

But though Magendie was a talented experimenter who kept scrupulous records, Bernard recognized a flaw in his teacher's methods. Trying to work around the vitalists, Magendie had thrown himself into experimental physiology with abandon—meaning in this case with little thought or plan of action. Bernard later wrote: "Magendie was in his manner an empiric, because he used to say: disturb the phenomena, that is, experiment and *simply verify without explaining anything.* That was still not a complete experimental method at all, because in the experimental method it is necessary to explain."[6]

In two areas, for example, Magendie both suggested experiments and monitored Bernard's work: on the study of gastric juices, and the experimental cutting of nerves. In both cases, however, as Bernard moved off on his own he carried the work to new and different levels.

Though Bernard had not shown any special talent as a student, under Magendie he flourished, and his new career was under way.

In 1844 he applied for, but did not get, a teaching post at the school. He resigned his position with Magendie and tried to set up his own small research laboratory and school of experimental physiology. The school went nowhere, and he had to carry out his experiments in a temporary lab of his own, or in those of a few of his colleagues in the chemical world. For chemists were respected and admired, and were given impressive laboratories for their work. The same held true for natural scientists such as botanists and zoologists. Experimental physiologists, however, were despised and reviled.

A Despised Profession

Even Magendie—a faculty member at the Collège de France, a haven for scientific inquiry—was relegated to a corner. Described as a damp and dark lair, it was, wrote Foster, "a hiding place for a wild beast." Magendie could offer no facilities for Bernard's own work.

Further, the apparent irregularity and unpredictability of physiological phenomena had led to a kind of defeatist attitude among other physiologists. Several of the leading practitioners simply left the field and switched to anatomy.[7]

Bernard spelled out this state of affairs in a later work of his: "So soon as an experimental physiologist was discovered he was denounced; he was given over to the reproaches of his neighbours and subjected to annoyances by the police."[8]

Around 1844, for example, Bernard was in the midst of studies on the digestive powers of the gastric juices. One of his experiments, being carried out in the laboratory of a chemist friend, involved use of a cannula, a sort of silver tap that could be turned on and off. This was fitted into the stomach of a live dog in such a way that samples of its gastric juices could be drawn off, while the health of the dog was in no way imperiled.

After the operation, the dog was let loose into the yard of the laboratory and was to be shown to a visiting German colleague the next morning. But the animal escaped during the night, carrying in its belly the "accusing cannula of a physiologist."[9]

A few days later Bernard was called into the police commissioner's office in the Rue du Jardinet; he was received very coldly and asked if he would admit to having placed the instrument seen in the belly of the dog, which the police now had on the premises. Bernard attempted to explain the situation, but this seemed only to anger the commissioner. One of his complaints had to do with the question of where Bernard had gotten the dog. (One scholar believes it was the commissioner's own dog, which had earlier disappeared.)[10] Bernard's explanation that it came from the very same men who were employed by the police to collect stray dogs seemed to help.

Bernard added that he regretted the pain suffered by the animal, and suggested a solution. He would remove the cannula, and leave the animal there, assuring the commissioner that the dog would be none the worse for the experience.

Visiting the office a few days later, he found the dog in good health, and the officer's attitude completely changed. Indeed they became friends, and Bernard could henceforth count on his protection. Bernard even set up his small laboratory in the commissioner's district, where he gave some private courses of experimental physiology under the license and protection of his new friend, "whereby I was saved many disagreeable incidents." [11]

Nevertheless, Bernard soon had to give up his quarters there. Even the sympathy of the friendly police commissioner was not enough to stem an antivivisectionist barrage of rumors. One of these was that children as well as animals were being brought clandestinely to his laboratory in sacks. [12]

The intensity of the antivivisection opposition was to plague him for the rest of his life. It was even to explode in his married life. During the difficult years between 1844 and 1847, Bernard was introduced to and married Marie François Martin, daughter of a Paris physician, who came along with a dowry. This enabled him to continue his researches on his own, but didn't leave them much to live on. Unfortunately, writes Foster, "she was not prepared, as he was, to live on narrow means, in order that the world might be the richer." [13] The life of a prosperous physician's wife would have been much more to her taste.

Though the marriage did not actually dissolve until 1870, his domestic life must have been a nightmare. Madame Bernard, with "narrow intelligence and a limited capacity for sympathy, was unable to share or understand her husband's intellectual life." [14] To make matters worse, she and one of their daughters spent much of their time and funds in antivivisectionist activities.

As happened over and over in his lifetime, he managed to bury himself in his work and carry on. A major change had occurred in 1847, however. Magendie was ailing and Bernard was appointed his deputy at the Collège. He thereby had an official laboratory, and over the next few years did some of his most brilliant work.

But even so, in spite of his powerful drive and his deep-seated belief that he was at the doorway to a new world of medical discovery, he despaired of the future. As Foster put it:

In 1851, at the very time that he was unlocking the gates of Fame, conscious though he must have been of the high value of the truths which he was beginning to make known, he was despairing of the future. The path of the experimental physiologist seemed so doubtful, the difficulties so many and so pressing, the hopes of success so few and so shadowy, that he at this time had serious thoughts of abandoning a career of science, and of devoting himself to active practice as a surgeon.[15]

Happily for us, he soldiered on.

Fruitful Years

Bernard's work over the years included, but was not limited to, studies of the physiology of the nervous system, the cerebrospinal fluid, the locus of oxidation in mammals (leading to the first real understanding of respiration and metabolism), and the physiology of digestion.

Over time, his remarkably diverse talents emerged, and he used a wide variety of techniques. In studying the working of nerves, he not only severed specific nerves and monitored the result, but also subjected them to galvanic stimulation.

He is given credit for being the first experimenter to perform cardiac catheterization.[16] To sample blood from a specific vein (the inferior vena cava) deep inside the animal's body, he managed to insert a tube via the jugular vein in the neck of the animal and through the heart.

He is also credited with being the first to keep alive an organ separated from the body by feeding it with needed substances (a technique called perfusion that is now widely used).[17]

Later descriptions of his work, by him and others, rarely show how difficult the work was, and how persistent he sometimes had to be. After one experiment in which he was trying to sample some blood immediately after it had passed through the liver, he wrote in his notebook: "It will be necessary to try with the tube to penetrate into the subhepatic veins, for that time perhaps I did not go far enough. Do it over."[18]

The broad reaches of his inquiries required that he also school himself well in chemical matters if he were not to be beholden to chemists who despised his work. He learned much from Justus von Liebig's writings and the lectures of a contemporary, J. B. A. Dumas. He also found a few chemists who were sympathetic to him; it was in the chemical laboratory of his friend T. J. Pelouze that he did some of his important work.

As had happened several times in his career, he turned a chance occurrence into a new scientific technique. In 1845, Pelouze had given him a sample of a substance he had brought back from the United States. This was curare, about which little was known at the time except that it was a potent poison.

In Bernard's hands, the substance provided yet another approach to vivisection, in this case chemical vivisection. Using it he found, for example, that curare causes death by destroying all the motor nerves, without affecting the sensory nerves. He thereby showed the difference between sensory nerves and those that control function.

This was an extremely important finding, one that has had many ramifications down the line. Curare works only where the nerve meets the muscle that it is controlling. This kind of specific action has been seen with other drugs and control points; thanks to Bernard's pioneering work in this area, the term "receptor" has become a major focus for drug researchers. Some breast cancer cells, for example, thrive on the female hormone estrogen. A recently developed drug, tamoxifen, binds to specific receptors on these cells, preventing the binding of estrogen and thereby stopping the growth of cancer cells.

Other materials he worked with included carbon monoxide, opium, strychnine, and anesthetics, which he used not only to deaden pain where possible, but as part of his experimental arsenal. In his hands, these substances provided a kind of analytical technique that could elucidate our subtlest functional systems. Through it one of his earlier assistants, Wilhelm Kühne, revealed the existence of that hitherto unsuspected structure, the neuromuscular junction, in 1862.[19]

Internal Environment

But the best part of his work was done when he reasoned his way to an important idea and then checked it out experimentally. A good example is his theory of the internal environment, which he developed over a period of 20 years, starting in the early 1850s. Worthy of a book itself, the basic idea is both simple and profound.

As Bernard himself put it: "All vital mechanisms, however varied they may be, have only one object, that of preserving constant the conditions of life in the internal environment."[20]

Bernard's idea was so revolutionary that when Michael Foster, himself an eminent physiologist, published his biography of Bernard in 1899, he made no mention of the *milieu intérieur*. But a generation later the equally eminent Scottish physiologist J. S. Haldane said of Bernard's words: "No more pregnant sentence was ever framed by a physiologist."[21]

The theory has had enormous consequences. In 1926 it led an American, Walter Cannon, to develop the idea of homeostasis, the process that keeps your body's systems in balance despite fluctuations in the environment. (Are your fluids depleted in hot, dry weather? You become thirsty and drink. You're not dressed warmly enough? You shiver and warm up.) It also led, later, to the modern concept of feedback, the fundamental basis for automated manufacturing, navigation, and all manner of modern technology.

In an astounding leap of insight, Bernard came up with two entirely different ideas to support his internal environment theory, both of which constitute fundamental, but quite different, methods of control in the body. One is "internal secretions," now commonly known as hormones, which are produced by the ductless glands; these perform a variety of biochemical functions as they circulate in the body. The second is nerve control of blood flow, an equally important bodily function.

These insights opened up a completely new conception of the regulation of such bodily phenomena as temperature and blood circulation; his work in the latter area continues today as the foundation for all studies of that 20th-century scourge, hypertension. He also seeded the important field now known as endocrinology.[22]

Diabetics who depend on the hormone insulin to balance out their blood sugar owe their lives to Bernard. The substance increases glucose uptake by muscle cells and increases the storage of glycogen and other such substances. It was only because of the broad range of his researches that he could begin to understand what was going on, thereby providing a basis for the later discovery of insulin itself. He was even able to create a case of diabetes in an experimental animal by cutting the appropriate nerves.

This was difficult work, for anyone and at any time, but especially in those unreceptive times. It required the mind of a theorist, the hands of a microsurgeon, the patience of a saint, and the tenaciousness of a bulldog. In his notes covering work spanning the years 1850–1860, he wrote, "Always pursue the idea that the physical and chemical phenomena are controlled by the nervous system."[23] It wasn't enough, however, for him to come up with this brilliant insight. He knew that he needed the supporting evidence as well. It was, after all, not even certain that one *could* analyze and understand these subtle, artful, and often transitory events.[24]

In Sickness and in Health

By the mid-1850s, Bernard began to think about consolidating some of his researches. He had achieved considerable recognition and had published many papers. It was in this period that he came up with his *milieu intérieur,* which he named in 1857. He began to think about producing a major work on his experimental method.

Then, in March 1860, his life took a severe turn. Years of intensive labor, often in less than perfect sanitary conditions, led to a major breakdown in his health. Leaving his wife and daughters in Paris, he retired to his boyhood home at Saint-Julien and lived alone in a country cottage. For some two and a half years he was virtually cut off from the outside professional world.

His illness was both painful and severe, possibly some sort of dysentery or colitis, perhaps a form of cholera. Bernard himself suspected that his health problems could be attributed to emotional disturbances.[25]

But it gave him time to think even more deeply, to begin pulling together the many ends he had uncovered. He also conceived of writing a major opus; it was to be a multivolume compendium of all he had learned in his many years of deep study.

As it turned out, all he managed to complete was the introduction to this work, but *Introduction à l'étude de la médecine expérimentale* (An introduction to the study of experimental medicine), originally published in 1865, has become a true science classic. In it he lays out a sensible, workable program for studies in experimental medicine. In the last section, he also illustrates the ideas using examples from his own studies, his reasoning being that "I shall be much more certain of what I describe in telling what has happened to me than in interpreting what may have taken place in the minds of others."

A modest man, he adds, "I am not, however, so fatuous as to give these examples as models to follow; I use them only to express my ideas better and to make my thoughts easier to grasp."[26]

He also saw clearly the potential of his method. "By a marvelous compensation, science, in humbling our pride, proportionately increases our power." Also: "We know absolutely nothing of the essence even of life, but we shall nevertheless regulate vital phenomena as soon as we know enough of their necessary conditions."[27] Bernard also threw in, along the way, some strong barbs aimed at the still active vitalistic doctrine.

In addition to being a true scientific classic, still well worth reading today for its medical insights, the *Introduction* is a delight to read. The same holds for his other writings, including many scientific papers. In fact, his earlier writing talent and experience paid off, and most of his influence on succeeding generations can be traced to his writings.

There was another, ironic, contributor to Bernard's impressive output, of both research and writings. One historian compares Bernard's life after 1865 with that of Carl Ludwig, a contemporary of Bernard's. Ludwig was set up in a well-appointed institute of physiology in Leipzig, and for 25 years was mentor to many young students. France, however, provided more honor than material support, and Bernard continued to be mostly an independent researcher, sometimes with assistance, but never as part of a real institute.[28] This, in some sense, may have worked in his favor; less burdened with administrative details—and by that time free of domestic obligations as well—he was more free to pursue his chosen activity.

A thoughtful and sensitive man, he recognized that his methods inevitably caused some pain and suffering in animals. "The science of life," he wrote in his *Introduction,* "is a superb and dazzling lighted hall which may be reached only by passing through a long and ghastly kitchen."[29]

The world around him concentrated more and more on the kitchen.

Opposition to Experimental Medicine

Other than at hospitals, 19th-century French medicine provided few opportunities for medical researchers, which is why Bernard had been so eager to move off on to his own. He later wrote: "I consider hospitals only as the entrance to scientific medicine . . . the true sanctuary of medical science is a laboratory. . . ."[30]

He felt, for example, that hospital medicine, and particularly as it was practiced in his country and his day, was strictly an observational science. It was passive. He believed that the phenomena being studied were too complex to be understood by just observing.

To overcome these disadvantages, it was necessary, he thought, to interrupt normal processes and to study them under carefully controlled conditions. Today it hardly seems credible, but in Bernard's day the very idea of experimental medicine still faced opposition from every quarter—from the medical establishment, from the chemical world, from other physiologists, and from an increasingly vocal group, the antivivisectionists. It may even have been the very successes of his approach that, by bringing attention to himself, brought also increased opposition.

Physicians

Bernard believed that "experimenters have to touch the body on which they wish to act, whether by destroying it or altering it, so as to learn the part which it plays in the phenomena of nature."[31] Many physicians, however, still believed it was beneath the dignity of a physician to dirty his hands with this sort of work. So even though Bernard had medical training, his medical colleagues provided little support.

They even found fault with his conclusions. For more than 50 years, for example, the British physician Frederick Pavy (1829–1911) did everything he could to discredit Bernard's theory that the liver both manufactured glycogen from other substances and also broke it down to supply sugar for the body's needs.

In spite of the fact that Pavy was wrong in practically every detail, he was constantly given good coverage in the medical journals.[32] A significant part of his confusion, and at the same time the reason for his success in publication, was his vitalistic view that molecules undergoing chemical change in the living organism were in some sense "alive."

Chemists

By Bernard's time, a chemical revolution had already taken place. Building on the successes of physical scientists like Galileo and Newton, chemists on the order of Priestley, Davy, and Dalton had all used experimental control in their chemical discoveries. The problem they had with Bernard, in other words, lay not in his use of experimental control, but in using it on living things.

Then, in 1828, the German chemist Friedrich Wöhler managed to synthesize urea, a substance long thought to be found only in living things. This especially seemed to support the belief of the few nonvitalist chemists that any process taking place in a living creature should be producible outside of it as well. Out of this came all sorts of strange and wonderful ideas.

One widespread theory was that valid conclusions could be drawn regarding the physiology of nutrition by monitoring what a creature took in, and comparing it with substances isolated from its tissues and from its excreta. Bernard had decided by 1848 that this was nonsense and argued against it. "It would be . . . like trying to tell what is taking place inside a house by watching what enters through the door and what leaves by the chimney." [33]

He didn't discount the possibility that physical and chemical actions might well be the same inside and outside the body. "But," he argued, "in the living being, they are also harmoniously organized and integrated, and this is their most striking characteristic." [34]

Living processes, in other words, and particularly those involving human life, were another matter altogether. Louis Mialhe became Bernard's prime example of one who took an excessively chemical approach to life's processes. Bernard challenged chemists "who reason from the laboratory to the organism, whereas it is necessary to reason from the organism to the laboratory (error of Mialhe)." [35] The long-running argument with Mialhe may have accentuated his irritation with chemists. [36]

Summarizing his approach later in life, he wrote: "Twenty-five years ago, at the beginning of my physiological career, I tried . . . to carry experimentation into the internal environment of the organism, in order to follow step by step and experimentally all of those transformations of materials which the chemists were explaining theoretically." [37]

Antivivisectionists

But if Bernard faced strong opposition from physicians and chemists, even more vehement was the direct response to the facilitating device in his work. For vivisection is a red flag that has long swung in the scientific breeze. Harvey, who carried out such experiments, was roundly excoriated for them.

Unfortunately for Bernard, some important changes had occurred in the intervening years, which led to even stronger opposition. The 18th century saw many severe abuses of animals, and some researchers may indeed have derived sadistic pleasure from their experiments. But an apparent obliviousness to animals' pain may also have derived from a widespread belief that animals, being less than humans in so many ways, do not feel pain. [38]

Nevertheless, there was already some backlash. Samuel Johnson, in his time England's most distinguished man of letters, complained in

1758 that reports of animal experiments were "being published every day with ostentation" by doctors who "extend the arts of torture."[39]

Still, there was in these earlier times an obvious connection between vivisection and medicine, which may have helped balance the outside world's revulsion against vivisection. By the 19th century, however, a kind of moral consciousness was erupting. Bernard himself may have contributed; though he always had medical applications in mind, the effect of his efforts was, at least initially, to separate physiology from medicine. This put his vivisection experiments in an even less-favorable light.

Growing Opposition

For a variety of reasons, the antivivisection eruption occurred first in England, where a powerful movement took root. In 1822, Parliament passed the first law concerning the mistreatment of animals. Two years later the Society for the Prevention of Cruelty to Animals was formed, a forerunner of many more to follow—on the Continent, in the United States, and elsewhere.

But these were generalized protection societies, and were mostly concerned with domesticated animals. It was not until 1870 that a movement specifically aimed at vivisection arose, again in England. And then the public jumped in with both feet. Other, even more stringent, laws followed that under certain circumstances made vivisection a crime.

There was, inevitably, a response to the response. Here's an interesting editorial, taken from an 1877 issue of the *Popular Science Monthly*.[40]

> The humane feelings of both sexes were profoundly stirred by the tales of atrocity that were circulated, so that the scientific physiologists of the country began to be looked upon as friends, reveling in the infliction of agony upon helpless animals. The stories, of course, were unscrupulous exaggerations, or arrant lies, but the public is a great believer and fond of pungent sensations while fervid philanthropy is not apt to trouble itself much about cool matters of evidence.

After some more description of the situation, in an equally ironic tone, the editorial continues:

That hysterical rampage of British philanthropy was strong enough to . . . extort from [the government] a law that was alike an insult to science and a disgrace to the country . . . it was legislated that he [the physiologist] should not pursue his work except by a license from a political office-holder, and the making of any experiment [that gives] pain to an animal was declared an offense punishable in the first instance by a fine, and in the second by fine and imprisonment. . . . In short, legislative wisdom, stimulated by philanthropic zeal, outlawed vivisection as a crime, and then provided for its perpetuation by leave of the Secretary of State.

But now comes the kicker. "Lords, Commoners, and everybody that can afford it, then seize their guns, and betake themselves to the fields and mountains wherever there is anything to be killed. . . . But most of them are bad shots, and wound many more than they kill. . . ." The author then refers to the pain and suffering this can cause.

There's more. Quoting from an article by a Mr. Lowe in the *Contemporary Review,* the editorial continues:

According to present British law . . . it appears that, while the man of science must not inflict the least pain on any animal for the most beneficent object, any one else may inflict the most exquisite tortures on any non-domestic animal—that is, on ninety-nine hundredths of brute creation—without any punishment at all. If he can show that the torture was inflicted from cruelty, from gluttony, for money, for amusement—for any motive, in fact, except a desire to do good by extending knowledge—he enjoys the most perfect impunity, but woe to him if in his infliction of pain there is any alloy of science.

Finally:

The agitation was . . . directed against a certain class of scientific men, and had its chief root in those narrow prejudices against science which the press and the pulpit have recently done so much to nourish and sustain. There has been an especial dread of biological science, because it meddles with the mysteries of life, and aims to explain things which ignorance and superstition would rather not have explained. . . . The anti-vivisection movement, in short, was very much a result of that feeling of jealousy and hostility toward science . . . an aggressive and intolerant movement.

This is the atmosphere in which Bernard was working, amid arguments that sound remarkably similar to those of today. Although the broadscale situation was less contentious in France than in England, it showed up there as well. As early as 1832, even before Bernard came on the scene, faculty members at the Collège de France were lodging official complaints against both vivisection and dissection. In fact, the most dangerous opposition came from scientists and physicians, probably for the reasons given above.

There were major objections outside the Collège as well. Even the great and dominant French naturalist Baron Georges Cuvier (1769–1832) mocked experiments on living animals, maintaining that they were simply sources of erroneous views.

Bernard's eminence, however, had a curious effect. Though it called attention to both him and his work, his reputation and position were such that the direct attacks were often aimed not at him but at Magendie, particularly after the latter's death in 1855. Starting in 1863, P. E. Chauffard, a professor at the Faculty of Medicine, and Dubois d'Amiens, secretary of the Academy of Medicine, actually accused Magendie of having carried out vivisection of living human beings. In response, Bernard called the objections "Don Quixotry."[41]

In 1882 the Société Française contre la Vivisection was formed, under the honorary chairmanship of the poet, novelist, and dramatist Victor Hugo. He was the ideal choice, not only for his preeminent position as a French man of letters, but also because of his known sympathy for all victims of social injustice—so eloquently expressed in his 1862 novel, *Les Misérables*.

Scientists, however, are no less human than practitioners of any kind, and many have been torn by the suffering of experimental animals. Charles Darwin, as gentle and sensitive a man as ever lived, was among those who had decidedly mixed feelings. In 1871 he wrote to Professor Ray Lankester, "You ask about my opinion on vivisection. I quite agree that it is justifiable for real investigations on physiology; but not for mere damnable and detestable curiosity. It is a subject which makes me sick with horror, so I will not say another word about it, else I shall not sleep tonight."[42]

But, 10 years later, he also wrote:

On the other hand, I know that physiology cannot possibly progress except by means of experiments on living animals, and I feel the deepest conviction that he who retards the progress of physiology commits a crime against mankind. Any one who remembers, as I can, the state of this science half a century ago must admit that it has made immense progress . . . no one, unless he is grossly ignorant of what science has done for mankind, can entertain any doubt of the incalculable benefits which will hereafter be derived from physiology, not only by man, but by the lower animals. . . . In the future everyone will be astonished at the ingratitude shown, at least in England, to these benefactors of mankind.[43]

Everyone? Well, maybe not everyone. The issue remains a highly contentious one,[44] with animal rights people no less exercised today than they were then.[45] Several militant groups, again in England, have actually put certain prominent scientists on a death list. One such scientist says he spends 20 percent of his time on defensive activities.[46] This, even though laws protecting experimental animals not only remain on the books but have even been strengthened, and even though the numbers of animals used has been declining for years.[47]

Death and Resurrection

Bernard, so far as we know, never had to face death threats. But he paid dearly in other ways. The break between Bernard and his family, for instance, was so complete that even as he lay dying, one of his daughters came to his apartment but would not go into his room. Sophie Raffalovich, a friend of Bernard's, later reported that she listened to his daughter's footsteps as she paced back and forth in an adjoining room and "the sound smote us to the heart."[48]

Upon his death in 1878, Bernard was accorded a great national funeral, the first French scientist to be so honored. But very quickly posterity—that fickle chooser of fame and recognition—forgot all about him. Perhaps it was just that his ideas were so revolutionary; perhaps it was the public's antipathy to his methods or his reluctance to act as his own publicity agent. Or it may have been the excitement generated by the newly emerging bacteriology of that other Frenchman, Louis Pasteur.

Today, while Pasteur's name is revered everywhere, the only members of the lay public who are likely to recognize Bernard's name are the animal rights people, to whom it is anathema.

To those working in the biomedical field, however, Bernard came alive again in our century, especially as new discoveries revealed just how groundbreaking his work had been. The man, his discoveries, and his methods all found new respect in the biomedical world.

Bernard didn't begin the practice of experimental medicine, but—virtually single-handedly and beleaguered on all sides—he turned it into a scientific discipline. We are all in his debt.

CHAPTER 5

Pasteur versus Liebig, Pouchet, and Koch

Fermentation, Spontaneous Generation, and Germ Theory

Prior to the mid-1880s, no physician had ever cured a disease. As a result the medical profession was not held in high esteem, and biting comments were common. Here's a sample from a satirical periodical called *Judge:* "It is to the credit of the intelligence of the medical profession that they do not often make the mistake of taking their own medicines."[1] And in another issue of the same periodical a month later: "Five times as many ambitious women take to medicine as to law. This contradicts that generally-received idea of the sex that they delight in scandals and quarrels, but abhor cruelty and killing."[2]

Yet at the same time, the press was beginning to receive, and was pumping up, a powerful new wave: namely, stories about treatments for serious diseases that actually seemed to work. The first to be reported concerned rabies, generally caused by the bite of a rabid animal. Until then, the only treatment available was branding of the bite area with a red-hot iron—which sometimes worked. Most often, the victim died a horrible, painful death anyway.

The new rabies treatment led to an attitudinal change of breathtaking proportions. Even in that backwater, the United States, the public's attitude toward the medical profession changed almost overnight from disdain to wonder. Doctors began to be looked at in a new light.

The excitement was so great, and the coverage so intensive, that the patients became heroes as well. Four young working-class American children who had been bitten by a rabid dog were sent to Paris to be treated by the new method. As the newspapers, both here and in Europe, reported the ongoing story, the boys' pictures and names were splashed across the front pages. Thousands of eager listeners in the United States later attended paid performances to hear the boys describe their experience.

Even more amazing, the name most often associated with this new development was not that of a kindly old doctor, but of Louis Pasteur, a chemist.

Pasteur had already acquired something of a reputation in several surprisingly varied areas—in basic chemistry for work on crystals; in the commercial world for saving the wine, beer, and silk industries; in the food production field for originating the process known as pasteurization; in the world of basic medical research for establishing and demonstrating the germ theory (that microorganisms lie at the heart of infectious disease); and in the livestock industry for developing vaccines against chicken cholera and also the virulent disease anthrax, which was devastating the country's cattle.

None of this, however, had to do directly with human clinical medicine. And, as I suggested earlier in this book, where human life is concerned, public interest leaps. That Pasteur's new treatment concerned rabies, a particularly vicious and frightening disease, and that the results were so obvious, made the cure seem almost miraculous.

The process—a series of about a dozen carefully prepared, increasingly virulent injections of the offending infectious material—was rigorous but clearly successful. Pasteur's name, already recognizable to several specific groups, suddenly became a household word. For his rabies treatment also raised the hope that other infectious diseases—typhus, typhoid, plague, cholera, diphtheria, syphilis, and many others—might now be both prevented and cured, which indeed has occurred in the years since then.

It's true that Pasteur's treatment was not the first use of inoculation. Edward Jenner had done it almost a century earlier with smallpox, but that was for prevention, not as a treatment for an already existing problem.

A medical revolution, in this case a kind of war in which everyone came out ahead, had begun.

The Most Perfect Man

One of Pasteur's early biographers called him "the most perfect man who has ever entered the kingdom of science . . . within the limits of humanity, well-nigh perfect."[3]

Paul de Kruif, a scientist and writer who knew Pasteur, called him "the scientific nonpareil, the microbe-hunting one and only . . ."[4]

The *Encyclopedia Americana* still says: "His discovery that most familiar diseases are caused by germs is the most important in medical history and one of the main foundations of modern medicine."[5]

More than a hundred research institutes now bear his name.[6]

All true. But this should not be taken to mean that his life path was smooth. A difficult man in many ways, he was not one to swallow criticism—and there was plenty to go around—without a fight.

Criticism

Reasons for all the criticism vary. It's true that Claude Bernard, a physician, laid the foundation for scientific medicine with his laboratory studies. But Pasteur, though a chemist, ended up laboring in the medical world itself. More than anyone else he showed that scientific research could pay off in the medical world. Still, as we have seen over and over, the medical world does not always take kindly to new developments, or to interlopers.

On the other hand, Pasteur was not always blameless. He lacked a sense of humor, and has been described by one biographer, Gerald L. Geison, as "profoundly serious, almost dour, and more than a little cool and aloof toward those outside his select circle."[7] All of these were characteristics that worked well in his professional life, but were hardly likely to endear him to his contemporaries.

Geison, a revisionist historian at Princeton University, adds: "In controversy, his combative self-assurance could be devastating to the point of cruelty. He so offended one opponent, an eighty-year-old surgeon, that the latter actually challenged him to a duel—which, happily for both, never took place."[8]

(Pasteurophiles see this skirmish differently, arguing that it took place at a scientific meeting, and that the challenger was reacting in a hotheaded manner more to Pasteur's revolutionary ideas than to his behavior.)

René Dubos, another of Pasteur's major biographers, wrote: "Pasteur became more and more inflammable as time went on. Not satisfied with challenging his opponents to disprove his claims, he heaped scorn upon their ignorance, their lack of experimental skill, their obtuseness or even their insincerity. . . ."[9]

At first, we wonder why he became so irascible. Though he suffered many tragedies—including loss of two beloved daughters and a sister to sickness—and had serious health problems by age 45, he had had good relations with his family, and with his early teachers and colleagues. As his name became better known, however, and as his interests and influence began to spread, opposition grew. And so did anger and frustration.

Thanks to this combustible combination—manifold interests and abilities, contentious personality, and complete faith in his abilities—he found himself involved in an amazing variety of wars, some quite intense. Like Bernard, he faced a powerful phalanx—including chemists, physicians, naturalists, and antivivisectionists.

All of this came later, however. His early career was quiet enough.

Crystals

Born and raised in a pastoral region of eastern France in 1822, he developed an interest in painting in his young life. His goal was to become a professor of fine arts. As part of his studies at the École Normale Supérieure in Paris, he attended lectures by the chemists Antoine-Jérôme Balard and, especially, Bernard's colleague Jean-Baptiste-André Dumas. Inspired by them, he decided to switch to chemistry.

His career got off to a good start. He received his Ph.D. in 1847 and only a year later had published a paper covering his doctoral work with crystalline materials. At least one reason Pasteur chose to work with crystals was their intrinsic beauty and intricate designs.

Through a series of brilliant and extremely delicate experiments, he discovered a highly significant fact, that molecules can be both "right-" and "left-handed." This led to a far more basic understanding of the chemicals contained in living tissue.

His finding gave him a jump-start up the ladder of success. By 1849, at the age of 27, he was professor of chemistry at the University of Strasbourg in northeastern France, where he taught and matured for five years. His love for chemistry emerged in his lectures, which became increasingly popular. It was in this period that he met and married Marie Laurent, who was to prove a steadfast, understanding mate through many difficult periods.

At the same time, his reputation was spreading and the quiet, contemplative young Pasteur was being slowly remodeled. He tasted the sweetness of flattery. Even the venerable French physicist Jean-Baptiste Biot praised him: "Pasteur throws light upon everything he touches." [10]

His talks at Paris's Academy of Sciences contributed significantly to France's scientific reputation. These lectures, incidentally, helped him develop and refine his public speaking skills, which were to prove indispensable in his later battles. In 1853 he was made a member of France's Legion of Honor, the first of many, many such honors.

A year later, at the still-young age of 32, his early work in crystallography earned him not only a professorship in chemistry at the University of Lille, but also the position of dean of the newly created Faculty of Sciences at the school.

Industrial Science

Lille was very different from the small towns in eastern France where Pasteur was born and brought up, and even from the still somewhat provincial city of Strasbourg. It was an important city in the booming industrial region of northern France, which included a strong presence in food production, breweries, and, especially, wineries. In fact, church authorities placed a new cathedral under the patronage of Notre Dame de la Treille (Our Lady of the Vineyard).

Pasteur's superiors had an interesting idea, one that was to exert a powerful influence on his later career: forget the idea of studying science for science's sake; apply your efforts toward obtaining useful results, results that can be directly applied in the region's industrial activities. Nor was this simply an academic exercise; several industries in the conurbation that was metropolitan Lille were in trouble and needed help. Perhaps Pasteur could make a difference.

This approach must have matched something in his own psyche, and he dove into his new administrative activities with skill and intelligence. He also honed his abilities at subtle diplomacy, often needed to attain desired objectives.

Results were excellent. He set up a successful academic program that attracted students who ended up able to deal with the practical realities of food, wine, and industrial production. Included were field trips to real plants. As he put it, students for too long had "approached the workers with ideas they have taken from novels."[11] Yet, at the same time, he made sure that theory was not ignored. What he was developing was in fact an early example of a technical university.

Busy as his schedule was, he somehow also found time to do some of his own research in a small laboratory he put together just one flight down from his own apartment.

Fermentation

During his inaugural speech he had asked, "Where . . . will you find a young man whose curiosity and interest will not immediately be aroused when you put into his hands a potato, when with that potato he may produce sugar, with that sugar alcohol, with that alcohol ether and vinegar?"[12]

He was here encapsulating the process called fermentation, which was to prove a major stepping-stone in his career. Think, for example, of the importance of bread and wine throughout history; both involve this process. Over thousands of years bakers and winemakers had plied their trade without any understanding of what was happening. The prevailing idea in the mid-19th century was that the fermentation process involved some sort of chemical change.

The same was thought to be true of the production of beer and cider, plus the souring of milk, as well as the spoilage of meat, eggs, and other foods. Pasteur's work with yeasts and bacteria provided the first understanding of fermentation and the things that can go wrong in it, along with the fact that living things lay at its heart. But the voyage was to turn into a long, hard one.

For he was going against established chemical doctrine. Of the objectors, the leader and standard-bearer was Baron Justus von Liebig, an arrogant, influential German chemist with a worldwide reputation. Liebig compared the biological idea of fermentation and putrefaction to "the opinion of a child who imagined that the rapid current of the Rhine depended on the movement of the many wheels on the mills on the Main River, which sent the waters towards the city of Bingen."[13] In other words, the biological explanation had everything backward.

As he saw it, fermentation was a sort of chemical decomposition. It was known that yeast, a microscopic plant, was involved, but he was sure it was the dead portion of the yeast that acted on the sugar. Decomposition is, of course, the opposite of life's processes, which are basically building and reproductive operations.

Liebig's was not the only theory that conflicted with Pasteur's, but it was a huge barrier that he had to overcome. For in truth Pasteur did not yet have a solid way of demonstrating his position, which he admitted. But he continued to delve deeper, to learn more, and to produce evidence in support of his position.

Liebig, though only 19 years older than Pasteur, was so old-fashioned that he even refused to look through a microscope. Pasteur put it to good use. It was he who first saw staphylococcus and pneumococcus bacteria. He saw that certain microorganisms not only do not require oxygen for life, but are killed by its presence. He considered this an important aspect of the fermentation process.

In response to the opposition, Pasteur put forth his experimental conclusions, but also criticized his opponents' methodology, which, he felt, was a throwback to the days of Scholasticism. In his long memoir of 1879, "The Physiological Theory of Fermentation," he wrote: "There is nothing more convenient than purely hypothetical theories, theories which are not the necessary consequences of facts; when fresh facts are discovered new hypotheses can be tacked on to the old ones. This is exactly what Liebig and Fremy have done, each in his turn, under the pressure of our studies, commenced in 1857." [14]

Pasteur began his memoir with the comment that his own theory dates back to 1843. Yet in 1879 he still felt it necessary to devote 23 pages of the paper to answering directly the objections and hypotheses of Liebig and his cohorts. Is it any wonder that his frustration level was rising?

Fermentation Gone Awry

One day in 1856, M. Bigo, a Lille manufacturer, visited Pasteur. He told the young dean that many of the regional producers, including himself, were producing alcohol from beetroot juice, and that they were experiencing major problems with both production and quality. Even the vats were giving off a sickening odor. The problems were threatening the entire industry.

Pasteur immersed himself in the problem and worked long and hard to find an answer. Often, when we read of scientists' successes, we lose track of the long roads involved, as when Pasteur wrote in his notebook, "mistake . . . all wrong . . . no . . ." [15] His wife, Marie, wrote to her father-in-law: "Louis is now up to his neck in beet juice. He spends his days in the alcohol factory." [16]

Yet at the same time, he never let go of his idea that fermentation was at heart a living process. Researchers had suggested much earlier that living yeast itself was somehow involved in the fermentation process, but they were always beaten down decisively by the opposition. Pasteur kept up the pressure.

In 1856–1857, in the midst of this struggle, he was sidetracked for a short time. Certain members of the influential Academy of Sciences in

Paris wanted to elect him to fill an opening in its very select membership. He was delighted and eager, but not optimistic. For very quickly, the same sort of opposition arose that was to haunt him throughout his long career.

According to one of his biographers, René Vallery-Radot (who happened also to be his admiring nephew), Pasteur "did not understand envy, ill-will, or jealousy, and was more than astonished, indeed amazed, when he came across such feelings." [17]

Still, Pasteur was realistic. "Everyone knows," he wrote to his wife, "that I am the valid candidate. . . . But they are afraid (at least many of them are) of chemistry. They are saying that chemistry wants to take over everything. That is why I have all the naturalists against me, especially the ignorant ones. . . ." [18]

He was not elected.

Another Move

It didn't matter. His reputation brought him another move up the ladder. He was appointed assistant director of both scientific studies and general administration at his old alma mater, the École Normale Supérieure, the same school he had attended not long before as an unexceptional student. So, at age 35, he moved to Paris with his wife and three small children, where he would spend the rest of his life.

But the exalted title was not matched by the trappings. He had neither laboratory nor research funds. Whatever little there was had been snatched by the chemistry department. Pasteur took over an attic room long abandoned to the rats and, using his own funds, equipped a small laboratory.

He went back to the fermentation problem, moved on to alcoholic fermentation, and, finally, published his *Mémoire sur la fermentation alcoolique* in 1860, in which he stated, flat out, that fermentation was *caused* by a living microorganism. Fermentation, he said, is the act of reproduction. And sugar, he pointed out, is the food of the reproducers.

As with all living processes, fermentation is a complicated one. He had to answer one of the strong objections to his idea, explaining why there might be less detectable yeast after fermentation than before. The nitrogen produced by dead yeasts, he argued, is used by the living ones for their own synthesis. In the process he was able to design a synthetic culture medium in which the yeast could flourish. Finally, within just a few years after M. Bigo had approached him, he had shown that a different kind of material, lactic acid, found in the brewers' vats was a contaminant, and that it had been produced by a different microorganism.

He also came up with a solution: heat the beet juice to destroy the contamination, then reseed with the appropriate alcoholic ferment.

By now, winemakers were also facing serious problems. So, in these cramped, unimpressive quarters, he completed in the mid-1860s studies begun at Lille, showing now that the wine problems, too, were being caused by one or more unwanted microorganisms. The solution, again, was heating. He had to show also that it was possible to heat the mixture enough to kill off unwanted microorganisms while not harming the flavor of the wine.

In 1866 he published a major book on diseases of wine and their prevention. It contained the first public description of what we now know as pasteurization.

He also discovered a surprising fact. Not only are some organisms killed by oxygen, but others, like yeast, can exist with or without it. In its absence the microscopic plant can operate without it, and causes fermentation in the process. He also showed that putrefaction, the negative side of fermentation from our point of view, is also caused by an anaerobic microbe, which he called a vibrio. Although putrefaction (rotting, decay, spoiling) may cause our nostrils to pucker up when we come across it, it is a perfectly normal process that helps keep our environment in balance by continually breaking down dead organic matter.

A lesser man might have let the situation rest there. Pasteur, however, simply used this work as a foundation for the next stage in his career. For, along the way, he began to wonder about the possible connections between microorganisms and disease in animals and humans. This should have moved him directly into the next major phase of his career, but he was continually deflected by the need to defend his ideas.

Implacable Opposition

His nemesis, for example, continued his implacable opposition. In 1869, Liebig was still standing tall and loud. Ten years after Pasteur had challenged him in print, Liebig finally came around to answering in print, but still held to his position. He did so in a long paper read to the

Royal Academy of Sciences in Munich, in which he also complained about difficulties in reproducing some of Pasteur's experiments.

Pasteur, outraged, quickly came back with a memoir in which he challenged the scientific world to decide between French biology and Liebig's German chemistry. He suggested that the Royal Academy appoint a commission that would judge the reproducibility and veracity of his experiments. Pasteur included in the challenge the very modern idea that they should together perform some new experiment, an experiment in which their respective ideas predicted contradictory results. Neither Liebig nor the academy responded.

Not one to give up easily, Pasteur traveled to Munich and practically forced a face-to-face meeting with Liebig. Liebig received him, but would not even ask him to be seated; he refused to discuss the matter, claiming that he did not feel well. Pasteur couldn't tell whether this aging, but still potent, opponent was really ill or was just too stubborn to give in.

Nor was Liebig his only opponent. Pasteur even had trouble with his friend and colleague Claude Bernard—after Bernard's death!—though Bernard would probably have been mortified if he had been alive to see how his name was used. He had, of course, been aware of the fermentation controversy, and had jotted down some notes. Among them were some that questioned (disputed, says Geison) Pasteur's germ theory of fermentation. Bernard never dreamed that his questions would see the light of day. He was simply mulling the theory in his own way.

One of Pasteur's major detractors, Marcelin Berthelot,[19] got hold of the notes and published them, thereby adding Bernard's good name to the list of those who opposed Pasteur. Pasteur was, of course, utterly distraught and was caught in a terrible bind. Bernard had been one of his dearest friends and supporters; each had provided professional and personal support to the other in time of need.

Pasteur felt he had to respond forcefully, or face yet another setback in his battle to get his ideas accepted. He issued a strongly worded critique, which in turn irritated a number of people—including, half a century later, Paul de Kruif, whose best-selling book *The Microbe Hunters* helped establish and popularize Pasteur's work in the United States. De Kruif complained that "this passionate Pasteur [had] jumped up and down on Bernard's grave."[20]

But, Pasteur argued, if his critique was damaging to Bernard's memory, then Berthelot must accept some of the blame.

We begin to understand why, of the many photographs we still see of Pasteur, there is only a single one that shows him smiling.

Ironically, Pasteur was not entirely right, and Liebig and his chemical compatriots were not entirely wrong. Follow-up work by Berthelot and others showed that fermentation can be accomplished by chemicals, which we now call enzymes. But it was not until 1897, two years after Pasteur's death, that the German chemist Eduard Buchner finally clarified the complex character of fermentation by extracting from yeast a chemical substance, which he called zymase, that could produce alcohol from sugar. He thereby showed definitively that living yeast cells produce a substance (a set of enzymes) that is not itself alive but that helps the chemical process along. In other words, as Pasteur suspected, living things are essential to the process at some point. His work also provided the foundation for today's enormously important world of enzymology.

In the midst of all this, he was already following up in other areas, including work in cell biology. And this led him into battle on not one but two fronts.

Spontaneous Generation

Since fermentation and putrescence (spoilage) are so common, and, in Pasteur's view, were caused by specific microorganisms, there were two logical conclusions that could be drawn: the germs are present everywhere and in very large numbers; or, they originate spontaneously in organic matter.

In this way he was brought face-to-face with spontaneous generation—the idea that life can spring from nonliving matter. Innumerable scholars and researchers had supported the idea for thousands of years. An experiment with flies in the 17th century by Francesco Redi seemed to lay it to rest. But with the arrival of the microscope, those who still believed in it could argue that even if flies don't arise spontaneously, that doesn't mean it can't happen in the world of microbes.

No definitive experiments had been done, and the general consensus was that if living tissue can turn into nonliving matter, why not the other way around? The most recent entrant in the debate was Félix Pouchet, director of the Natural History Museum in Rouen and a member of many scientific societies. He had published a paper in 1858 and an impressive book in 1859, both in support of the doctrine, but with a twist.

What Pouchet had in mind was a kind of reorganization of matter that had once been alive—by some sort of *force plastique*. He didn't spell that one out. He did point out, however, that the new creatures would not likely have parents like themselves, which led to his defining term, "heterogenesis."

To his credit, he did try to get an answer via experimentation. Basically, he filled a flask with boiling water that presumably was germfree, then sealed it shut. After immersing it completely in a trough of mercury, he let it cool; then he opened it under the mercury, while adding oxygen and a solution of common hay, which had also been boiled. The hay, he argued, would also start out germfree and the mercury would keep it that way.

Upon microscopic examination later, microorganisms did appear. This showed clearly, he maintained, that spontaneous generation was indeed a reality.

The French Academy of Sciences became interested and in January 1860 offered a prize to anyone who could throw new light on the question. Pasteur saw this as an irresistible challenge. His friends Dumas and Biot tried to steer him away; Biot warned: "You will never find your way out."[21]

As usual, Pasteur went his own way. Looking back into the history of the subject, he read about a set of experiments by John Turberville Needham in the mid-1700s that seemed to support the idea; and then an answering set by Lazzaro Spallanzani that refuted it.[22]

Then Pasteur sprang into action. In a series of brilliant experiments he showed that the microorganisms in Pouchet's experiment did not arise spontaneously, but were carried in by dust particles that had settled in the mercury itself.

In his own experiments he used the often described "swan-necked" flasks. These were flasks with long, curving necks that were left open, but in which dust would settle at the low point of the neck and never reach the sterile solutions.

In 1862, Pouchet withdrew from the competition and Pasteur was awarded the prize.

Nevertheless, Pouchet published a new book delineating his views and his experiment in 1864. In it he expressed exasperation at Pasteur's obstinacy in the face of "facts."[23] To the modern mind, his conception of facts was based more on appeal to authority and to speculation than on experimental method.

There was some more back-and-forth maneuvering on both sides. Then, on April 7, 1864, Pasteur presented a major address outlining the controversy and its outcome. He began by commenting: "Enough of poetry . . . , enough of fantasy and instinctive solutions. It is time that science, the true method, reclaims its rights and exercises them."[24]

Then he got down to business and delivered a careful description of his experiments. Finally, he expressed confidence that "never again will the doctrine of spontaneous generation recover from the mortal blow dealt by this simple experiment."[25]

His expression "simple experiment," by the way, has to be the understatement of the century. In one phase of his experiments, answering a new challenge by Pouchet, he worked with 60 flasks: 20 opened at the foot of a mountain, 20 at about 2,500 feet up, and the balance at the top of a glacier at 6,000 feet. The point was to show that it is possible to obtain air at any location that will not cause the spontaneous generation of microorganisms.

Silkworms

After this adventure, Pasteur turned back to his studies on fermentation and putrefaction. He was therefore the logical person to consult when the silkworm raisers in southern France suddenly found themselves faced with an epidemic of some sort that threatened to destroy the industry entirely. After thousands of years in which they turned silkworms into gold, they now found that all the phases of worm production—eggs, worms, pupae, and moths—were affected by the disease. No one had any idea what was wrong or what to do.

Starting with no knowledge whatever about silkworms, Pasteur immersed himself in the world of silkworms and the new threat. Again, virtually single-handedly, he saved a whole industry. It took several years, however, for he had to understand first that the worms were being attacked by not one but two different diseases, and two different microorganisms.

But his unremitting toil—plus, perhaps, the constant criticism from numerous sources—took its toll, and in 1868, at age 46, he was stricken by a paralysis so severe that no one expected him to survive. He did, but, as Victor Robinson writes: "Some of his final observations on the silkworm problem were dictated while the paralyzed man sat in an arm-chair with pupils around him. Slowly, the paralysis lifted, but never entirely, and for years to come, one of the greatest figures in modern science moved through laboratories with a stiffened hand and limping foot."[26]

And once again the foundation he built was to lead him to new heights, and a new controversy.

Germ Theory of Disease

Ever since 1857, when Pasteur had published his first paper on fermentation, he had been playing with the idea that there is a connection between germs and disease. His work with the diseases in silkworms pushed him even further in this direction. He worked continually to spread his germ theory of disease—but with little success.

For example, one day in the late 1870s a famous physician was holding forth at the Academy of Medicine in Paris. Using lots of Greek and Latin terminology, he was explaining childbed fever as some sort of metabolic disorder.

Suddenly a voice bellowed from the rear of the hall: "The thing that kills women with childbed fever—it isn't anything like that! It is you doctors who carry deadly microbes from sick women to healthy ones . . . !"[27] The voice was Pasteur's, and he was again embroiled in one of his many celebrated battles.

For despite the solid insights of Semmelweis, set forth a quarter of a century earlier (chapter 3), the medical establishment still refused to accept contagion as the disease's cause. Pasteur, however, had something that Semmelweis did not, a mechanism to explain what was happening.

The idea behind his germ theory seems simple enough today. Microorganisms such as bacteria, fungi, and protozoan parasites come from forebears of the same species; they are present in large numbers almost everywhere—in air, in water, in dust; finally, germs can therefore be understood as the causative factor in not only fermentation and putrescence but also in disease. Different diseases are caused by different microbes. It all fits together into such a neat package that it's hard at first to understand why his germ theory was so controversial and took so long to be accepted.

Recall, however, how difficult it had been earlier, and still was in Pasteur's day, to combat vitalistic beliefs in the essential difference between living material and inanimate matter. There was still enough confusion that his ideas could be conflated with vitalism, and were sometimes used by its proponents to support their position. To Pasteur's opponents, appealing to a living "germ" to explain a chemical reaction was a step backward.

Also, try to put yourself into the mind of someone alive at that time. Does it really make sense to think that a living thing as tiny as a bacterium could do the damage we see in severe diseases of animals and humans? Even the microscope, which Pasteur used to good effect, was a highly suspect instrument (witness Liebig).

But, severe as had been the criticisms of Pasteur by naturalists and chemists, Victor Robinson writes, "it was as nothing compared with the enmity which greeted him when he came among the doctors."[28] Who was this chemist to tell respected physicians that they did not know what they were talking about? "When Pasteur presented facts," Robinson continues, "their answer was: Monsieur, where is your M.D.? The leading academicians, men who are still remembered by historians, Chassaignac, Piorry, Pidoux, denounced the germ theory of disease in the most scathing and ironical language of which Frenchmen are capable. Today their words read like chapters out of a mediaeval book."[29] Among the charges leveled at Pasteur were "microbial madness," "fanaticism for the microbe," and "fetishism."[30]

Pasteur did not take all this lying down, of course. "What you lack, M. Frémy, is familiarity with the microbe, and you, M. Trécul, are not accustomed to laboratories."[31]

Yet at the very same time that physicians were arguing against the idea that germs were in some way related to infections, a bacterium was killing off cattle throughout France in massive numbers. What had been a standoff for centuries was turning into a catastrophe in France.

Again, this versatile chemist waged war, now against *Bacillus anthracis,* the anthrax bacterium. His major weapon was an attenuated (weakened) vaccine of the type that he had a few years before developed to deal—successfully—with chicken cholera.

But the anthrax bacillus has an extremely sturdy resting phase that can survive many challenges and for long periods while still retaining its power to cause disease. This made the battle more complex, and gave more ground for opposition. Nevertheless, by 1881 he had a vaccine for anthrax.

He set up a powerful, large-scale public demonstration of the vaccine's abilities at Pouilly-le-Fort, a farm just outside of Paris. It provided convincing evidence to livestock people of its value and they quickly put it to use. The medical profession, however, including veterinarians, remained skeptical.

He was now also besieged by two additional groups. One was the "antivaccinators," who advanced all kinds of objections. More important were the antivivisectionists, who were unhappy with his use of animals in his large-scale experiments and demonstrations (which ended up in the death of the animals used as controls). In fact, after Bernard's death in 1878, Pasteur became the chief target of the French antivivisection movement. As his fame spilled over into other parts of Europe, and especially into England, he found himself under attack from that region, too.

In the face of all this, Pasteur was already applying the vaccine idea[32] to human disease, with the happy results reported at the beginning of this chapter. Yet claims still arose that the rabies vaccine was useless and even dangerous. A major claim was that other factors were at work in the supposed "cures."

Even as terrified patients with rabies—real or suspected—jammed his office, the attacks on Pasteur continued. He had to endure a long, bitter battle with another antagonist: Robert Koch, a German research physician, who had also been working on contagious diseases. No scientific lightweight, Koch's publications in 1878 and 1879 had helped confirm that bacteria were the cause, and not the consequence, of infection.

Some of the polemics, on both sides, stemmed from a deep enmity between the German and French nations. Pasteur had been deeply affected when the Germans trounced France in the Franco-Prussian War of 1870–1871; he even returned an honorary degree that had been awarded him by a German university. He made no secret of the fact that his major objective in working with French beer brewers in the 1870s was to challenge the vaunted superiority of German beers.

His strong nationalism, in fact, helped make him a loved figure in France. After all, what French patriot would not respond with tears to his comment that he wanted to use his science to cure the "Prussian tumor"?[33] Koch, on his part, sneered at French microbiology at the same time as he voiced scathing doubts about the purity of the vaccine.

During all this, Pasteur managed to continue his experiments, devising many that answered one objection after another. Again, however, his health began to suffer; he now showed signs of heart trouble.

Eventually, as we know, he emerged victorious, in terms of both the anthrax and rabies inoculations; his rabies treatment was soon used to treat thousands of rabies victims. The income from the treatment was used to found the Pasteur Institute in Paris, which he directed for the last eight years of his life—even though he became increasingly feeble as the years passed.

As he prophesied, vaccines have now been developed for many diseases, including typhoid fever, which had carried off two of his beloved daughters. His work provided convincing evidence for both the theory and the practice of vaccination. Although it also built a foundation for the later development of antibiotics and other useful medicines, vaccines remain the most effective means of controlling infectious diseases.

Pasteur's feelings about the Germans never changed. In one of his last acts, he refused to accept the Prussian Order of Merit. A couple of months later, in September 1895, his various ailments finally caught up with him. He died quietly at his home in Villeneuve-l'Étang, just outside of Paris, surrounded by family and few of his disciples.

His funeral, an impressive state affair, was attended by France's president and officials from countries as distant as Greece and Russia.

End of story? Not quite.

A Modern Contretemps

Something about Pasteur must invite controversy. Starting around the beginning of the 1990s, Gerald L. Geison, mentioned earlier in this chapter, mounted an intensive study of Pasteur's 102 laboratory notebooks, which have been made available only in the last quarter century.[34] Based on these studies, he has produced an impressive, carefully detailed 378-page book, *The Private Science of Louis Pasteur.*

He argues in it that "the standard Pastorian legend needs to be qualified, even transformed," and that Pasteur's success owes much to his "polemical virtuosity and political savvy." He warns that as you read, "the standard Pastorian saga will begin to unravel."[35]

He maintains that "the word 'deception' no longer seems so inappropriate, and even 'fraud' does not seem entirely out of line."[36]

Geison has chosen to focus on a few episodes in Pasteur's career where there are "distinct—and sometimes astonishing—discrepancies between the results reported in his published papers and those recorded in his private manuscripts."[37]

For example, Geison points out that during the feud with Pouchet, Pasteur defined any experiment in which life did appear as "unsuccessful."[38] Another example he uses was not unknown before Geison's research, but here he spells it out in great detail.[39] Basically, it has to do with the fact that in the large-scale public trial of Pasteur's anthrax vaccine at Pouilly-le-Fort, he actually used a different vaccine than the one he had developed and that everyone assumed he was using.

In such cases, however, we enter a real quagmire: the inevitable differences between the actual work and original notations of the scientist, versus the public record of that work. I talked about this briefly in the chapter on Bernard.

Geison himself states, "In every case thus far in which records of 'private science' have been closely investigated, one can detect discrepancies of one sort or another between these records and published accounts." He adds, "Even the best scientists routinely dismiss uncongenial data as aberrations, arising from 'bad runs,' and therefore omit or 'suppress' them from the public record. . . ."[40]

This, Geison admits, is what might be called ordinary scientific method. Yet, he argues early in his book, "it should gradually become clear that some of Pasteur's most important work often failed to conform to ordinary notions of proper Scientific Method."[41] Among these are the spontaneous generation and anthrax researches. He argues, in other words, that Pasteur's actions fall outside of this "ordinary" scientific method.

Is Geison merely shooting off his mouth? Hardly. This is a careful, meticulously researched book. Most of the dozen reviews I turned up admired Geison's scholarship and tended to side with his point of view. There were, however, some interesting objections.

Bernard Dixon, a well-known writer and editor in the biotechnology field, charges: "Denigration of once-admired figures is now far more popular than hagiography."[42]

Roy Porter, an eminent British medical historian, feels that Geison "seems to write less as a historian than as one responding to the contemporary American obsession with medical ethics." He feels the revelations "do little to change" the broad outlines of the Pasteur story.[43]

Another objection has to do with what I call selective perception. One of Geison's charges against Pasteur is that he "sometimes clung tenaciously to 'preconceived ideas' even in the face of powerful evidence against them."[44] Elizabeth Fee, a professor of hygiene and public health at Johns Hopkins University, wrote a very positive review of the book, but feels nevertheless that "at times Geison seems to emphasize the negative, selecting the most unfavorable interpretation of Pasteur's actions when a kinder view may also be plausible."[45]

Perhaps this lay at the heart of one reviewer's outraged response. Max F. Perutz, a world-renowned British biochemist, seeks to deconstruct Geison's deconstruction. In his very long and detailed essay review in the *New York Review of Books,* he argues, in essence, "Some of these claims are scientifically flawed, while others defy common sense."[46]

But now the complexities mount, with layer upon layer of enigma, and we even begin to get involved with the two writers' inclinations. Geison maintains that his intentions are not to destroy Pasteur's name: "Let it be clear at the outset that I am less concerned to *expose* Pasteur's public deceptions than to *explain* them."[47]

Perutz, however, charges that Geison "follows the line laid down by certain social theorists who assert that scientific results are relative and subjective, because scientists interpret empirical facts in the light of their political and religious beliefs . . ."[48]—in other words, that Geison has taken Pasteur to task on grounds other than the purely scientific and has a specific agenda that he wishes to accomplish.

All of Pasteur's story makes for good reading (see the Bibliography). But if you really like a good fight, try reading or at least skimming Geison's book, then Perutz's review in the *New York Review of Books,* and then, in a later issue of the same publication, Geison's answer, and then Perutz's response to Geison's answer.[49] Here you will see clearly how different people can look at the same facts and come up with diametrically opposed conclusions. It is a sobering experience, and should help you keep your emotions under control when you read about, say, a nutritional claim today—and its exact opposite next week.[50]

Geison's book certainly helps humanize the process by which science takes place, which I am all for. But at the same time, it does much to smudge a shining image of a real medical hero. This, I think, may in the long run do more harm than good. We have few enough scientific heroes in the annals of history; I hate to see any of them deconstructed.

As one reviewer put it, "Most reviewers are saying now that Mr. Geison's book presents a measured and subtle analysis. It should not prompt sensationalist headlines, they say, but nevertheless it is doing so." After one of Geison's lectures on the topic, one report was headlined: "How Pasteur Perfected Science of Cheating."[51] Unfortunately, this is likely to be what the public will remember.

Considering the many years of stubborn obstinacy and fierce feuding that Pasteur endured, it's no wonder that he bent the facts somewhat to strengthen his case. It would have taken a saint, or a fool, to do otherwise. He was neither.

CHAPTER 6

Golgi versus Ramón y Cajal

The Nerve Network

"How distressing it is to have continually to fight against other men to defend the truth, instead of fighting against nature to wrest new truths from it! But how can it be avoided? Who does not know that every scientific accomplishment dislodges some deeply rooted error and that behind it is usually concealed injured pride, if not enraged interest?" [1]

Written early in this century by Santiago Ramón y Cajal, one of the truly great medical researchers of our time, this sad plaint reflects a life of struggle and disappointment. Although Cajal's story has a happy ending in a professional sense, decades of unremitting toil, of continued frustration and setbacks, took its toll. By his late 40s he "was attacked by neurasthenia, with palpitations, cardiac irregularities, insomnia, etc., with the resulting mental depression."[2] Earlier, he had suffered from both malaria and tuberculosis.

And who were these "other men" Cajal referred to? Mainly they were histologists, as he was—researchers interested in how the microscopical anatomy of living tissue relates to its function. If these "other men" had been a club of established medical researchers, then Camillo Golgi would have been its president.

What is especially sad is that both Cajal and Golgi were physicians and were hoping that what they learned could be applied to both mental and physical disorders. Too, they were working in an area that was, to put it simply, a deep, dark secret. So if any mistakes were made, it was to be expected. But the mistake that Golgi made was a big one, and it held back the field's progress for years.

That mistake, the "deeply rooted error" Cajal refers to, has to do with the neuron network of humans and other living things, surely one of the most complex structures on the earth. From the work of Galvani (chapter 2), researchers knew that electrical impulses were part of the process. But in the mid-1800s no one could even answer the basic question, How is the nervous impulse transmitted?

One problem was that the optical microscopes of the time were not much more useful than a strong magnifying glass. To make matters worse, the untreated brain tissue they were working with had the consistency of wet Kleenex, and had to be hardened chemically before it could be examined. But even when hardened, sliced, and put on the optical stage, their samples still looked like a mass of fresh-cooked, uncolored gelatin.

In an alternative method of study, technicians used fine needles and tried to free the minute cells from the surrounding tissue. Cajal called it an "undertaking for a Benedictine."[3]

Nevertheless, earlier researchers had determined that there were cells in the nervous system, and that they were different from other cells that had been seen in the body: nerve cells seemed to be sprouting shoots of some sort. But all that could be seen were the very beginnings of the shoots.

The general belief at the time was that all of these cells and processes (both short and extended fibers) actually constituted a continuous, interconnected web. This was not an unreasonable idea. It helped explain the remarkable speed with which thought and action take place.

The Golgi Stain

Then, in 1873—the same year in which Cajal got his medical degree—Golgi developed a laboratory method that enabled him and other histologists to get their first good look at the nervous system. By this time, chemical methods of hardening brain tissue had been developed. After many attempts, Golgi came up with a method that involved prolonged immersion of hardened brain tissue in a solution of silver nitrate—the same chemical solution that provides the darkening of grains in the photographic developing process.

Placing a thin slice of the block of tissue under the microscope, he saw a mass of dark blobs lying within a tangle of fibers. There, for the first time, were the neural cells and their varying processes, or shoots, outlined strongly—like India ink drawings—against a clear, translucent background.

Equally remarkable, the other cells did not darken, as they tended to with other stains in use for studying cells. This meant that the ones that did darken stood out even more clearly, without interference from all the others.

What Golgi saw enabled him to provide the clearest description yet of what has come to be called the neuron. He saw that the heart of the neuron was the cell, a tiny, massy blob that could take any of several different shapes, as had been seen in other types of cell. But what distinguished these were tiny shoots, or branches, which he was able to see clearly for the first time. We know now that they are of two kinds: a long, thin process called an axon; and a much shorter, highly branching set, called dendrites. Golgi even described two types of axons, with short and long extensions.

Reticular Theory

The Golgi stain was a remarkable technique and opened up the field for further study, by him and others as well. Unfortunately, Golgi's observations led him to support, and finally to become the leading advocate of, the idea that the entire nervous system in humans and other vertebrates was one solid network of fibers.

Called the reticular hypothesis, the idea was not new. It had been advanced two decades earlier and was supported by most of the neurologists of the day. But Golgi had "proved it" with his new staining method and the power of his fine reputation.

The only problem was that he was wrong. Ironically, although Golgi published some important work as a result of his observations, he began to lose faith in his own staining method. He also seems to have seen, and/or understood, less in his stained sections than Cajal later did. Golgi, for example, saw tiny spines extending from certain of the dendrites; but he ignored them, believing that the dendrites themselves served some sort of nutritive function and so were of little importance in signal transmission. Cajal, using similar equipment, described them very carefully and, as we'll see later, these spines turn out to have an important function in nerve transmission.

Was Golgi blind to things that should have been obvious? Hardly. The interpretation of these early slide sections called for some real leaps of interpretation, indeed of faith. That, however, didn't stop Golgi's compatriots from leaping aboard his bandwagon and championing his apparently solid demonstration of the truth of the reticular hypothesis.

Golgi and his colleagues had plenty of time to change their minds.

Cajal didn't learn of the procedure until 1887, four years after Golgi's discovery, by which time the reticular hypothesis had had time to "set."

Cajal quickly found ways to improve on the Golgi stain. He found, for example, that it only worked well with brain tissue that was, as he put it, "absolutely fresh, almost alive." He made another major discovery: embryonic nervous tissue produced far better results. For as the embryo matured, the longer processes (axons) begin to be encased in a fatty covering called myelin that obscures these important parts of the network.

Whereas Golgi essentially abandoned his own stain,[4] Cajal stuck with it. Although he made no major discoveries at first, he persisted and, finally, used it to disprove Golgi's dictum. The new truth, he said, "rose up in my mind like a revelation."[5]

By 1888 his observations convinced him that the nervous system actually consists of individual nerve cells, that the cells are the principal constituent, and that the various parts do indeed contact each other but do not connect physically. He described the Golgi-type network as "a sort of unfathomable physiologic sea." And he added that the reticular hypothesis, "by pretending to explain everything, explains absolutely nothing."

As was the case with Golgi's idea, Cajal's neural hypothesis had been propounded earlier, but only as vague suggestions. Cajal's observations clinched the idea. And so Cajal was quickly recognized as a major player on the world's scientific stage, and everyone lived happily ever after.

Well, not exactly.

For Golgi and the other holders of the original position still held the reins, and Cajal's early papers simply never appeared, even though he had given full credit to Golgi for his work.

Cajal finally realized that he was shouting in the wilderness—the wilderness being the world of Spanish science. French and German were the languages of science; all important publications and lectures were in these two languages. Not a single important scientist of Cajal's day could speak or understand Spanish, the only language Cajal knew.

Round Two

Cajal realized, finally, that to be heard he would have to do something drastic. He crafted a plan.

To be sure that his papers would appear, he founded a journal in

1888 and began to publish his own work. One of his publications, produced at considerable expense, had a print run of 60 copies. He sent copies off to various professors, but the papers were either ignored or were contemptuously dismissed. Still, he finally had some of his work in print.

Then he turned his attention abroad. He began to learn French and, fighting his inclination to work alone, he joined the German Anatomical Society.

Finally, and most importantly, in 1889 he packed up his evidence—carefully prepared sections of cerebellum, retina, and spinal cord—and went out with it to a meeting of the society to "sell" his idea to his contemporaries. His first real break came when Albert von Kölliker, the patriarch of German histology, saw the value of these remarkable preparations, took Cajal under his wing, and began to spread the word.

This, more than anything else, eventually turned the tide. But not easily, and not quickly. Although Cajal made a few additional converts, he found himself facing an even more deep-rooted problem. As he explained, "It was admitted [in the scientific world] that Spain might have produced some artist of genius, some long-haired poet, the gesticulating dancers of both sexes; but the supposition that a real man of science could arise within her was considered absurd."[6]

But Cajal continued to work furiously and successfully. He was awarded some major prizes, and delivered some major lectures (in his still-weak French).

It was during one of these lectures, the Croonian at the Royal Society of London in 1894, that Cajal introduced another of his major ideas, which he called dynamic polarity, and which is today accepted as a well-established law. Basically, it says that the direction of the nervous impulse is one-way, from dendrite to cell body to axon. Most contemporary workers in the field doubted that the dendrites had any electrical function at all.

After Cajal's lecture, another speaker suggested that "thanks to the work of Cajal, the impenetrable forest had been converted into a regular and pleasant park."[7]

Invited to the United States, he delivered a set of lectures at Clark University. These were published in 1899 and, with 31 beautiful illustrations, are still considered an important contribution to the subject. An interesting sidelight: In his third Clark lecture, Cajal pointed out his feeling that the United States "seems to be wonderfully endowed to triumph in the arena of scientific research."[8] Remember, this was at a time when France and Germany were still the premier bastions of scientific research.

A little later, in 1905, he was able to throw some light on the still puzzling neuronic ganglia, masses of nerve cells found in various parts of the body, especially the brain.

The Nobel Prize

Then, in 1906, Cajal was awarded the highly coveted Nobel Prize, to be shared with Golgi for their work on the nervous system. Perhaps the organizers were not aware of the antipathy between the two men; perhaps they didn't care. The result, however, was embarrassing in the extreme. Totally ignoring Cajal's contributions, Golgi still pushed the reticular hypothesis. He also mentioned the dendritic spines, which he had disregarded earlier, again ignoring Cajal.

Cajal later wrote, "What a cruel irony of fate to pair, like Siamese twins united by the shoulders, scientific adversaries of such contrasting character." Then he added, "I have never understood those strange mental constitutions which are devoted throughout life to the worship of their own egos, hermetically sealed to all innovation and impermeable to the incessant changes taking place in the intellectual environment."[9]

Still, Cajal was now standing on the Nobel stage, clearly recognized as a world-class scientist. Now, surely, his ideas and reputation were secure.

Well, not exactly.

Does fame provide a bulletproof vest? Cajal later wrote that the prize gave him more fear than pleasure. In his autobiography he asked himself,

How would my foreign opponents take the gifts of my lucky star? What would all those scientists, whose errors I had had the misfortune to show up, say about me?

I was not long in finding out. [A] few histologists and naturalists who always distinguished me with their disdain or their unfriendliness rose violently against my modest person. It was high time, according to my pious confrères, to crush the neuron doctrine for good, burying at the same time its most fervent supporter. There was in their invectives so much injustice, they were accompanied by such virulent personalities, and they [the invectives] were, finally, so disproportionate to the insignificance of my polite observations of earlier times, that it would be ingenuous to believe that there was not a certain etiological connection between them and the award of the Nobel Prize.[10]

In other words, the fight was not yet over. Even his old friend, Professor H. Held—who had earlier been an ardent supporter of Cajal's neuron doctrine, and had translated one of Cajal's books into German—turned against him.[11]

As we know, Cajal did in fact prevail. Slowly the tide rose in his favor and eclipsed the continuous (reticular) theory entirely—but with an interesting postscript, to be added later. In the years from about 1886 to 1906, it would be fair to say that Cajal laid the foundations for the modern interpretation of the nervous system and, virtually single-handedly, established a recognized and respected Spanish school of neuroscience.

Round Three

Thus far, the points have all gone to Cajal. It's the story one usually hears or reads about these two men and their part in the untangling of the neural net. But there's more to the story than that. In retrospect, Cajal has an advantage, for he wrote voluminously of his personal life. And a fascinating life it was. One writer has used his autobiography as the basis for a novel based on Cajal's growing up years.[12]

Too, he comes across in his own writings as a highly thoughtful, sympathetic, and sensitive person. In describing the passage of nervous impulses from one neuron to another, he wrote of "protoplasmic kisses . . . which seem to constitute the final ecstasy of an epic love story."[13]

Was he truly all goodness and light? He states in his autobiography, for example, that his main interest was in encouraging his followers. But when one of his workers published his own paper on glial, or supporting, cells, Cajal dismissed him, though they were later reconciled.[14] Still, we must grant Cajal this: Even after the Nobel contretemps, he could still write, "Golgi was a master of his method of anatomical analysis and completed our knowledge of the neuron."[15]

And how about Golgi? Was he personally as bad as he has been painted, by others as well as Cajal?

A deeper understanding of the story, and of the feud, requires that we learn something about the two men themselves.

Ramón y Cajal

Cajal was born in May 1852 in the small, bleak Spanish town of Navarre. His father, Justo Ramón Casasús, was a "modest surgeon" (Cajal's term), meaning he was not a full physician, but was permitted to do minor operations. Justo later, by dint of untiring effort, became a full-fledged physician. Cajal described his father as "a man of great energy, an extraordinarily hard worker, and full of noble ambition." [16]

The two defining characteristics of Cajal's early years were an astonishing rebelliousness—relatives described him as "wilful and unbearable"—and an irresistible urge to become an artist. Justo tried desperately to beat the idea out of him. He wanted Santiago to become a doctor.

Cajal, trying later to understand and come to terms with the beatings and deprivations he suffered, wrote, "There is actually in the function of the teacher something of the arrogant satisfaction of the breaker of colts. . . ." But Cajal recognized another side in his father as well, and wrote of "the kindly curiosity of the gardener who eagerly awaits the spring to find out the colour of the flower he has sown and to test the success of his methods of cultivation." [17]

When, at first, Justo thought that his ambition for his son was hopeless, when Santiago wouldn't or couldn't seem to do anything right at school, Justo apprenticed the young man to a barber and then to a boot maker.

All to no avail. Even at the boot maker, Santiago found ways to make use of his artistic abilities. And although always in trouble—with his teachers, his employers, or his own father—he somehow always found a way to practice his art, even when his father beat him and snatched away whatever pencils or brushes he was able to borrow, craft, or steal.

But Justo was by now a respected physician and surgeon, and apparently a good one. He also finally recognized that his son had a special talent in art. Cleverly, he managed to get Santiago interested in anatomy by working with him as they, together, studied the bones and other parts of bodies that they dug up from a neighborhood cemetery!

Now, suddenly, Santiago's artistic talents found a place, for he began to produce some of the finest anatomical drawings his father—and later, the world—had ever seen. It reflected a major talent that Santiago had, an ability to really see things that others would miss, and it stood him in good stead later. For histology probably makes greater use of and depends more on illustration than any other branch of medicine. Inaccurate, or incomplete, such illustrations can do more harm than good when used in medical texts. His drawings became the standard against which all others had to be compared.

His fighting spirit, too, was part of Cajal, and was needed in his long fight for recognition. It was honed by experiences at home and at a variety of strict schools that would have squelched a lesser man. Throughout all his travails, however, his deep love of country never changed. He gave as one reason for going into medicine: "When, in my youth, I saw with sadness how greatly anatomy and biology had deteriorated in Spain and how few of my fellow countrymen had won a place in the history of scientific medicine, I made a firm resolution to abandon forever my artistic ambitions, the golden dream of my youth, and to sally forth boldly into the international lists of biological research." [18]

Though he began by studying medicine—receiving his degree from the University of Zaragoza in 1873—his love of anatomy became almost an obsession and drew him inexorably into research. The long hours he spent peering at the dark shapes and tangles were leavened by the mysterious beauty he saw in them.

Remember, too, that in this period (the 1870s), "many, perhaps the majority of professors . . . despised the microscope, considering it even prejudicial to the progress of Biology!" [19]

With so little solid information available on microscopic vertebrate anatomy, he proceeded to work his way through the entire realm of microscopic anatomy. Then, "there came the turn of the nervous system, that masterpiece of life. I examined it eagerly in various animals." [20]

Part of the reason for his concentration in this area was that he, like his father, had noble, even grandiose, ambitions. Cajal felt that if he could tease apart the anatomy of the nervous system, perhaps he could also elucidate some of the mysteries of thought, maybe even find answers to some of the problems of mental (nervous) illness.

And now, finally, his inventive genius began to flower. He published many books and papers, his ability to illustrate his own publications so effectively being a strong reason for the world to finally recognize his contributions. But even then he didn't ease up.

He regarded his investigations into ganglia from 1903 to 1913 as among the "most fortunate" of his scientific work. [21] In the latter part of that period he made important contributions in the understanding of neuroglia, or glial cells. Earlier thought to be merely a supporting skeleton, they actually serve an important physiological function for the active nerve elements.

Golgi

And then we have Camillo Golgi. Far less is known of his early years. Born in Brescia, Italy, in 1843 he, too, came from a medical family, and he, too, became a physician—graduating in 1865 from the prestigious University of Pavia. He worked for a short time in a psychiatric clinic, but already his mind and heart were turning toward the anatomy and pathology of the nervous system, mainly via microscopic studies of nerve tissue. His first publications appeared in 1868 and 1871, and by 1873 he had formulated a full-fledged theory of the nervous system. It is unfortunate that his name is so often connected with the reticular fiasco, for much of his work was very solid and created a good foundation for Cajal to build on.

In 1879 he was appointed to the chair of anatomy at the highly respected University of Siena, after which he returned to his old stamping grounds at Pavia, which he built into a major center of research on the nervous system. Also, during the years 1885–1893, he did important work on malaria.

His contributions by no means stop there. He worked also in the field of cytology—the study of cells—and in 1898 was the first to show the existence of a fascinating subdivision within the cell. What he saw and described—an amazing accomplishment considering the equipment he was working with—was the existence within nerve cells of a small organ, or organelle, in the shape of a network of flattened, interlaced threads.

Irony

Poor Golgi. The organelle that he described, which came to be called the Golgi apparatus, fell under a cloud of suspicion. For almost half a century histologists questioned its very existence. As recently as the late 1940s, George E. Palade, who later shared a Nobel Prize for his work in cytology, still saw the Golgi apparatus as an artifact of the staining process.

Again, it took the development of more modern and more powerful techniques to show that the Golgi apparatus is not only real but also an important constituent of a cell. One powerful combination of techniques involves radioactive labeling, staining, and electron microscopy. What would Golgi and Cajal have given for access to this powerful tool chest!

Only in recent years, in fact, have histologists been able to tease out the secrets of the organelle. In the 1960s, George E. Palade—the same Palade who had questioned its existence two decades earlier—showed that proteins important in the functioning of the cell pass through the Golgi apparatus.

Then, starting around 1980, the Golgi apparatus was seen to have a precise internal structure. James E. Rothman, who has done important work in elucidating the organelle's activities, recently described them as "tiny membrane-enclosed sacs that are flattened and stacked like dinner plates."[22]

Further work showed that the apparatus contains an array of enzymes that carry out several important functions. In the almost infinite complexity of our cells, some 10,000 different proteins are involved, each needing to find its way to someplace else. The Golgi apparatus has been found to be an important protein finishing and distribution center.[23]

Researchers have also found cases in which Golgi's reticular idea hold true. In certain regions of the central nervous system, the pathway *is* continuous! Susan A. Greenfield, a professor of pharmacology and medicine at Oxford University, explains:

> If neurons were fused together and worked just by conducting electrical impulses, then it would be much faster. As it happens, there are some neuron-to-neuron contacts where the neurons appear fused together and there is no need for a chemical synapse [the commonly accepted explanation for how the individual nerve cells pass the nerve impulses from one to another]. Ironically, Golgi was, at least in these cases, correct after all.[24]

Contributions

So, all in all, we needn't feel sorry for him. He remained a highly respected member of the scientific establishment. Such prominent members were sometimes rewarded with political appointments and, around 1900, Golgi became a senator and began to play an active role in both the academic and public worlds. Perhaps the trappings of power were somehow at the heart of his inability to deal gracefully with what eventually was seen to be a major error on his part.

Among histologists and neuroscientists, however, his name is still revered. At the 1973 Golgi Centennial Symposium, celebrating the 100th anniversary of his initial discovery, histologists gathered to commemorate his life and work. The symposium proceedings were published in 1975.

Dominick P. Purpura, who kicked off the event, pointed out:

Despite the fact that the Golgi method is generally considered one of the most unpredictable neurohistological techniques, it remains today the single most important method for studying the structural organization of the brain. Evidently, capriciousness is happily tolerated for the esthetic reward of neurons revealed in all their majesty. There can be no question that the Golgi method will continue to be the standard against which other methods will be compared for some time to come.[25]

He added: "Indeed, adequate three-dimensional analysis of Golgi impregnated neurons has been possible only recently." Referring to computer-assisted methods that combine Golgi and electron microscope techniques, he termed them "neogolgi" methods.[26]

Nor have even these new methods closed out the field. Purpura added: "Many problems of concern nearly a century ago continue to attract our attention today." Among them is the role that synaptic transmission plays in complex behaviors, particularly learning and memory. It's interesting to note that "no coherent theory or unifying principle of neurobiology and behavior has emerged as yet. . . . In effect the Golgi method provided the first and most comprehensive analytical tool for this type of investigation."[27]

Finally, Purpura heralded "a return of the Golgi method to neuropathology after nearly a half century of total abandonment."[28] J. Szentágothai suggested, however, that "the temporary loss in drive and impetus during the 20s, 30s and 40s of histology was perhaps a necessity" in that it gave physiologists (who study the functions and activities of living matter) a chance to catch up with anatomists. The pause also made it possible for histologists "to realize that it was the *reazione nera* [the black reaction] of Golgi that gave the most realistic picture of neuronal structures. . . ."[29]

Of course, Cajal has his 20th-century admirers as well. In 1930, Dr. William H. F. Addison, of the Graduate School of Medicine, University of Pennsylvania, wrote an "Appreciation," in which he pointed out, "Following him, many have investigated special parts of the brain, and time and again have found themselves merely elaborating what Professor Cajal had already observed."[30]

And more recently, Dr. Gregory K. Bergey, reviewing a new English translation of Cajal's epic *Histology of the Nervous System of Man and Vertebrates* for the *New England Journal of Medicine*, compares Cajal's development of the concept of the neuron doctrine to William Harvey's contributions to our understanding of blood circulation, and his *Histology of the Nervous System* to Darwin's *Origin of the Species*.[31] Finally, some of his slides were sent up in the 1998 Neurolab space shuttle mission as part of a research project.

Yet for years, both before and after Cajal's death, his neuron doctrine was frequently attributed to someone else entirely, mainly to Wilhelm von Waldeyer, a German anatomist. Cajal deeply resented this, for Waldeyer had never worked in this field, though he had written a summary of Cajal's ideas, and had coined the term *neurone*[32] to describe the nerve cell and its branches.

By World War I, Cajal's reputation had fallen to the point where a Spanish translation of a popular French book used in Spanish medical schools mentioned Golgi and Waldeyer, but not Cajal![33]

Happily, the reputations of both Cajal and Golgi now seem secure.

In trying to evaluate the feud, we have to wonder why Golgi behaved as he did. Certainly, scientists who do important work like to be recognized. Certainly, a Nobel Prize is important recognition. But why did sharing it with Cajal seem to bother Golgi so much?[34] Was he just being pigheaded? Was he upset that Cajal, who had made his name on Golgi's technique, had outshown him?[35] Did he really continue to believe that he was right and Cajal was wrong? It's quite possible, for at the time, remember, the reticular hypothesis was seeing a strong resurgence.

We may never know. It is just possible, however, that in holding Cajal down in his earlier years, Golgi and his followers forced Cajal to delve even deeper into the subject than he would have otherwise, and drove him to come up with even stronger evidence.

Today

Modern techniques have shown clearly how nerve impulses travel, using both electrical and chemical routes. Cajal's dynamic polarity (one-way transmission) has been clearly substantiated. Yet as complex and devious a device the central nervous system is, there are exceptions to the one-way rule. In some instances the impulses do indeed travel in both directions.

Professor Greenfield summarizes what we now know about the brain:

> The brain is built up from single neurons in increasingly complex circuits. Between ten thousand and one hundred thousand neurons make contact with any particular neuron. In turn, any particular neuron will become one of many thousands of inputs for the next cell in the network. If we took a piece of brain the size of a match head alone, there could be up to a billion connections on that surface.[36]

Is it any wonder that Golgi made a mistake?

As for why he behaved as he did, perhaps his behavior just gives evidence, in case we needed it, that he was a human being as well as a scientist.

CHAPTER 7

Freud versus Moll, Breuer, Jung, and Many Others

Psychoanalysis

In that powerful thriller *The Silence of the Lambs,* Clarice Starling, an FBI trainee with some psychological background, attempts to enlist the aid of a vicious, imprisoned psychopath, Hannibal Lecter, in a desperate attempt to track down a cunning serial murderer. Hannibal teases Clarice, yet provides just enough real help—in the form of subtle psychological insights about the murderer, nicknamed Buffalo Bill—to keep her coming back for more.

But at the same time Hannibal, a trained psychiatrist, looks deep into Clarice's soul, and draws out of her, as a quid pro quo, details of a troubled childhood; he gradually learns that her father was murdered and that she spent some of her growing-up period at a ranch that butchered lambs. Lecter asks her, "Do you think if you caught Buffalo Bill yourself . . . you could make the lambs stop screaming, do you think they'd be all right too and you wouldn't wake up again in the dark and hear the lambs screaming?"

It's a powerful image—Clarice's nightmares and their connection with some buried horror in her background. Hopeful, uncertain, Clarice answers, "Yes. I don't know. Maybe." [1]

Would the book, which was made into a spine-tingling movie, have had the same kick if Sigmund Freud had never existed? Would it even have been written?

A Tidal Wave

At the turn of the 21st century, Freudians are celebrating the 100th anniversary of *The Interpretation of Dreams,* Freud's seminal work. He was certainly not the first to write about dreams, but, as in so much else that he did, he took what had been guesses—for example, that dreams were auguries of the future, or were just "a kind of spasm occurring in a mind that is otherwise asleep"[2]—and built on them to create an important, usable method for inquiring into our minds and souls.

Freud also wrote about crime, and some of his insights have indeed been used in attempts to fathom the inclinations, needs, and desires of criminals.[3] For example, he saw an interesting correlation between criminals and neurotics. In both cases, he pointed out, the investigator is attempting to unearth something hidden. But the criminal takes great pains to conceal the hidden activity or information from the outside world, while the neurotic is concealing something from himself. In addition, although there is resistance in both cases, the neurotic is clearly trying, with some part of his brain, to cooperate.

Freud, in fact, saw interesting correlations in an astonishing range of subjects. For although he started out his professional career as a physician and neurologist, as he delved deeper and deeper into the hidden crevasses of his own and other minds, he slowly began to move away from traditional biomedical treatments of mental illness.

It's worth looking at these treatments for a moment, to give a better idea of the medical scene in Freud's day. Because madness was generally thought of as an organic disease, physicians—whether alienists, psychiatrists, or asylum keepers—treated it as a medical condition: they bled the patients; they used opiates to sedate or brandy to stimulate; they purged the system with laxatives and emetics; and they used rest, massage, hot and cold baths, and a variety of mostly ineffectual electrotherapies (see chapter 2).

At first Freud thought he could create some sort of biopsychotherapeutic science, one based in the biomedical science of the time (combining both psychological and biological aspects), but he slowly came to feel that there had to be something better than traditional treatments. As he moved toward a more open-ended psychological approach, however, he came in contact with a far wider range of thoughts and activities.

There was, perhaps, an even more important factor leading to his wide range of interests; he felt that he did not have a true medical temperament, that he was missing (one can see the twinkle in his eyes) the "innate sadistic disposition."[4] And so, as Philip Rieff puts it, he made himself over into a new kind of doctor, one who could thereby "claim all society as his patient. . . . Freud's [kind of] physician was to be a student of history, religion, and the arts."[5]

During Freud's long and productive life—he died in 1939 at the age of 83—he wrote in all of those areas. Consider the diversity of a few of his books: *Jokes and Their Relation to the Unconscious* (1905); *Totem and Taboo* (1913), *Civilization and Its Discontents* (1930), and *Moses and Monotheism* (1939). He also produced a vast number of essays. Interested in the unconscious sources of literature, he wrote a charming and insightful analysis of a novella by Danish writer Johannes Vilhelm Jensen, which he called "Delusions and Dreams in Jensen's *Gradiva*" (1907); he did the first psychoanalytical biography, using Leonardo da Vinci as his subject (1910); and he produced a revealing essay on Michelangelo's statue of Moses (1912).

He was, of course, interested in the pathological fantasies of neurotics, but also in what he saw as the fantasies of whole nations as seen in their various legends, myths, folklore, and customs, which opened up whole additional worlds to him.

Others have written in these many areas, too,[6] but, as in so many other ways, he changed our way of thinking in every one of them.

And all of this was aside from his basic ideas and contributions in the worlds of psychoanalysis, sexuality, and consciousness. Harold Bloom, widely recognized as one of America's leading literary critics, has written that Freud "is at once the principal writer and the principal thinker of our century." He adds: "Perhaps his effect upon us is even more important than the apparently lasting value of his general theory of the mind."[7]

When one confronts the considerable corpus of Freud's own writings, plus the massive literature on both him and his ideas,[8] there is no escaping the fact that the man was simply a force of nature. Freud's influence is so basic, so widespread, that one recent scholar, Dr. Ernest Hartmann, a professor of psychiatry at Tufts University School of Medicine, describes his influence as a tidal wave. As Hartmann succinctly put it in a recent lecture:

> In our brave new world
> You can rave and rant
> But you can't
> Avoid Freud.[9]

Depending on your point of view, this is either admirable/awesome or terrifying/detestable.

Destined for Conflict

Freud, like Moses, gathered a group of faithful followers and took them into uncharted, unfriendly territory. For Moses and the Israelites—

wandering in a desolate region, surrounded on all sides by hostile forces—those were indeed perilous times. Freud, however, made us all face an even more frightening world—our own inner maelstrom, which had lain hidden from earliest times.

Are we, for example, not in control of our own thoughts? Are we not rational beings? As Freud saw it,

> Psychoanalysis has concluded from a study of the dreams and mental slips of normal people, as well as from the symptoms of neurotics, that the primitive, savage and evil impulses of mankind have not vanished in any individual, but continue their existence, although in a repressed state—in the unconscious, as we call it in our language—and they wait for opportunities to display their activity.
>
> It has further taught us that our intellect is a feeble and dependent thing, a plaything and tool of our impulses and emotions; that all of us are forced to behave cleverly or stupidly according as our attitudes and inner resistances ordain.[10]

At least as troubling were his ideas on the centrality of sex and the sex drive and, even worse, infant sexuality. Young children, even infants, are not only capable of experiencing erotic sensations but, he argued, unresolved conflicts in infantile sexuality[11] are prime shapers of personality and a strong driver in neurosis.

Other writers had touched on these ideas. Freud himself noted that such thinkers as Nietzsche and Schopenhauer had offered some interesting insights. Schopenhauer, for example, argued for the existence of unconscious mental processes, and even attributed insanity to some sort of sexual repression.

And still others, some less well known to us but highly respected in Freud's day, also saw significance in the sexual world. Among them were Albert Moll, Leopold Löwenfeld, Albert von Schrenck-Notzing, Richard von Krafft-Ebing, and Havelock Ellis. Krafft-Ebing's notorious *Psychopathia Sexualis* (1886) turned a strong spotlight on the till-then dark world of sexual perversions, and may even have prepared some ground for Freud's work.[12] But in general, Freud maintained, he preferred to develop his ideas on his own and only read about the others' later.

But where those researchers and writers had cast out some disparate ideas, Freud built on them to create full-fledged, in-your-face concepts, especially when considered in the light of his newly developing method of probing the mind, which he called psychoanalysis. As a result, their ideas never had the incendiary effect that Freud's did. As just one example, Karl Kraus, a satirist and contemporary of Freud's, maintained that Freud's psychoanalysis was the mental illness, not the cure.[13]

Especially interesting are cases in which colleagues and admirers became antagonists and detractors.

From Collaboration to Collision

Freud's relationship with Albert Moll, a Berlin physician and one of those named above, is particularly interesting. He had a lot in common with Freud: his father was also a Jewish merchant; like Freud, he delved into hypnosis, sexuality, and the psychotherapeutic treatment of sexual perversions. And he died on the very same day as did Freud— September 24, 1939.

Yet although his name is little known today, his career moved considerably faster than Freud's. His first book, on hypnotism (1889), led to worldwide recognition, and by the turn of the century he was a well-known neurologist and possibly the premier authority on sexual pathology.

At first he and Freud were friendly and corresponded happily. But they basically had very different ideas on the subject of infantile sex and the part it plays in later neuroses; the difference grew into a long, drawn-out brawl.

In 1905, Freud's book *Three Essays on the Theory of Sexuality* was published. Ernest Jones, an early Freudian and one of Freud's major biographers, writes of its reception, "*The Interpretation of Dreams* had been hailed as fantastic and ridiculous, but the *Three Essays* were shockingly wicked. Freud was a man with an evil and obscene mind. . . . This assault on the pristine innocence of childhood was unforgivable." Also: "The book certainly brought down on him more odium than any of his other writings."[14]

Among the attackers was the same Albert Moll, who issued a strong criticism of Freud's *Three Essays* in his own 1908 book, *The Sexual Life of Children*.[15] Though he didn't question the existence of infant sexuality, he, along with British psychologist Havelock Ellis, argued that it was normal and not a cause of later problems. He therefore felt that Freud's method of studying the sexual life of children via the memories of adults, rather than by investigating their sexual lives directly, was just plain wrong.[16] And he didn't hesitate to strongly criticize the whole idea of psychoanalysis.

Freud responded strongly. At a group discussion of Moll's book at the November 11, 1908, meeting of the Vienna Psycho-Analytical Society, Freud maintained that "Moll gleaned the importance of infantile sexuality from the *Three Essays,* and then proceeded to write this book. . . . Moll's whole book is permeated by the desire to deny Freud's influence."[17]

The dispute had turned into a priority spat, with each seeking the mantle of discoverer of infantile sexuality. Freud argued, and Jones seconds the idea, that Moll had earlier (before publication of Freud's *Three Essays*) *denied* the existence of infantile sexuality. Jones writes of Moll's book: "It was so vehement in denial of infantile sexuality that Freud said in a letter to a colleague, 'There are several passages that would justify a libel action, but silence is the best answer.' "[18]

The battle with Moll ran on for years. But it was a curious one, with Moll trying to behave as if there was no problem. In 1909 he visited Freud, who, however, felt Moll was behaving in a patronizing manner. At this Freud lost his temper and the meeting ended badly.

In a letter to his then close friend Carl Jung, Freud wrote: "To put it bluntly, he is a brute . . . a pettifogging lawyer. I was amazed to discover that he regards himself as a kind of patron of our movement. . . . He had stunk up the room like the devil himself. . . . Now of course we can expect all kinds of dirty tricks from him."[19]

Moll, in fact, had already argued in his book, and he continued to argue,

> The impression produced in my mind is that the theory of Freud and his followers suffices to account for the clinical histories, not that the clinical histories suffice to prove the truth of the theory. Freud endeavors to establish his theory by the aid of psychoanalysis. But this involves so many arbitrary interpretations, that it is impossible to speak of proof in any strict sense of the term.[20]

(This is an argument that has dogged Freud and his ideas right through our own era.)

By 1914, Freud was convinced that a society Moll had started in Berlin was intended mainly as a way of achieving recognition for Wilhelm Fliess, another colleague who had had an even closer tie with

Freud and whose relationship with him had also ended in estrangement and hostility. Finally, in 1926, Freud withdrew from a group that was preparing for a forthcoming meeting of the Congress on Sex Research when he found out that Moll was to lead the congress.

Who really came up with the idea of infant sexuality? It may not be possible to sort out the details. Several later writers argue that Jones has misconstrued some of what happened in Freud's day, at least partly because of his own experiences—he himself lost his job in England for asking a young girl about her sexual thoughts and experiences while treating her, and had to move to Canada.

Frank J. Sulloway writes in his extensive biography: "Freud had a particularly important reason for wanting to be seen as the discoverer of infantile sexuality. He knew that this discovery constituted a prime piece of scientific propaganda in favor of the importance and efficacy of his new method of exploring the mind. . . ."[21]

The likelihood is that the same explanation will serve here as we have heard earlier. The specific idea may not have been Freud's. But it is what he did with it that made his work special, and that brought it to the fore.

But Moll and Fliess are only two of many of Freud's contemporaries who began as colleagues and/or friends and ended as detractors and even enemies. A possible explanation: He and the small group he gathered around him at first were, or at least felt, isolated from the established medical world, including psychiatric circles. This led to a kind of wagon-circle-defense mentality and drove the group's members closer together. But as often happens in circles that are too close, explosion and fire are common results. And it couldn't have helped that many of those who gravitated toward this explorer and his work tended to be lonely and/or troubled themselves.

Another source of conflict was that some of his followers were not only very bright but very aggressive. Freud feared that if these followers were allowed to claim the mantle of psychoanalysis while going off, sometimes noisily, in various directions, this could derail his own attempts at creating a unified method.

Outside the immediate circle, few felt comfortable with his ideas; many were outraged. All in all, Freud was treading in dangerous waters. Of the feuders considered in this book, in fact, none has been involved in both the amount and the intensity of controversy that Freud and his ideas have inspired. It was widespread during his life, and grew as his ideas continued to spread after his death in 1939.

Or so Freud and his followers would have us believe. We'll consider these claims again later in this chapter. But before we do, it would be well (and in keeping with the subject) to take a closer look at his early years.

Early Development

Freud entered the University of Vienna's medical school in 1873, at the age of 17. Even then, he was more interested in studying human health, both mental and physical, than in learning about treating disease. His constant delving into biological researches caused his school career to stretch three years longer than usual—he finally graduated as a physician in 1881—but it gave him a strong grounding for subsequent studies of the human brain and the nervous system.

He spent the years 1882 to 1885 as a neurologist and psychiatrist at Vienna's famous Allgemeines Krankenhaus (General Hospital), at which, in truth, conditions had not improved greatly since Semmelweis's experience there. In 1883 he experimented with a new process he had developed for hardening and staining sections of the brain in preparation for microscopic examination.[22]

Freud had to struggle to make ends meet during his early professional years; he even put off his marriage for several years because he didn't feel he could support a family on his meager earnings. In any case, he decided to put aside his researches for a while and set up a clinical practice. This clinical experience played a major role in his later researches.

At first, he used the same sort of mostly ineffectual treatments as everyone else in the field was using. But he soon suspected that any successes, as with the electrotherapy, could be explained in other ways. In his autobiography he later explained, "So I put my electrical apparatus aside, even before Möbius [explained] that the successes of electrical treatment (in so far as there were any) were the effect of suggestion on the part of the physician."[23]

He began to look in other directions. By the mid-1880s, Freud had become interested in hysteria, a common affliction in his day, and was intrigued by the renowned physiologist Josef Breuer's treatment of it by hypnosis. In 1885 he traveled to Paris and studied the method with French neurologist Jean-Martin Charcot; he was deeply impressed by Charcot's ability to create contraction and even paralysis in limbs by hypnosis, the very same symptoms he was seeing in some "hysterical" patients, and by some of the cures he saw Charcot accomplish.

By 1891, Freud had produced three works on cerebral paralysis in children, and had brought some order to what had been a confusing mixture of paralyses. He took up Breuer's "cathartic[24] method," and they collaborated on a case that has come down to us as "Anna O." They also collaborated on a major book, *Studies in Hysteria,* published in 1895. It outlined their "talking cure"[25] and is often taken as the beginning of psychoanalysis.

Though this sounds like Breuer should be given credit for "inventing" psychoanalysis, there were major differences. The cathartic cure was a suggestive one; the physician did most of the talking. This is very different from the approach Freud was to take in his psychoanalysis—listening carefully to a patient's ramblings, including memories and dreams. It also depended on hypnosis, which meant that those patients who couldn't be hypnotized couldn't be treated. Finally, Freud was troubled by what he saw as a kind of tyranny in Bruer's approach—the doctor might shout at the patient, "What are you doing?!" He saw this as "an evident injustice and violence."[26]

Finally, nowhere in their book was sex singled out over other possible factors in a patient's background. Freud was already seeing things differently, and as he talked and wrote up his changing ideas on the importance of sexual etiologies, the storm that began to brew frightened Breuer, and he quickly backed away from both his collaborator and from the field. Breuer had built up a good medical practice, and had no desire to be pulled down with his colleague.

Freud and Breuer had, prior to the latter's retreat, become very close, both professionally and personally. Breuer had even lent Freud money. Jones says that Freud was by this time supporting a dozen people, not including servants, and that Breuer had advanced him a considerable sum.[27]

As happened over and over again with other colleagues, Freud saw Breuer's withdrawal not only as a desertion, but as a significant double cross, for he felt desperately in need of supporters and companions, and he took it hard. "The development of psychoanalysis," he wrote later, ". . . cost me his friendship. It was not easy for me to pay such a price, but I could not escape it."[28]

Freud's drive and determination showed itself, however. As upset as he was—at one point when they met, Breuer greeted Freud, but Freud chose to cut him—he simply moved on to the next phase in his development. In the summer of 1897 he began his own self-analysis—there were, after all, no other psychoanalysts to go to. It was a brave and difficult task, particularly in light of what was apparently a rather strong puritanical streak in his own personality—which, à la Freud, suggests all kinds of interesting ideas. By the fall he had converted the Oedipus myth into the Oedipus complex[29] and began serious work on infantile sexuality.

The Oedipus complex could certainly have had connections with his own life. Freud had always been his mother's "golden Siggie"; she foresaw big things for him, and her expectations became his. His relationship with his father, however, was very different, and led to other less salubrious, but just as important, results. Among them was an apparent continuous search for a satisfactory father figure. One aspect of his own father that troubled him greatly was Jakob's passive acceptance of anti-Semitic insults.

So the great Sigmund Freud, the originator and then leader of a towering movement, spent a significant part of his life and energy in a debilitating search for a satisfactory father figure.

The result of Freud's search was a series of extremely close personal/professional alliances, which too often ended disastrously. It happened with mentors such as Breuer and Fliess, and also with a series of followers, including both Alfred Adler and Carl Jung, both of whom went on to found their own institutes. The break with Jung was particularly painful for him.

An Emotional Break

In 1902 a select group of devoted followers began meeting on Wednesday evenings at Freud's house. Whether from too much delving into the unconscious or for other reasons, Freud by this time—even though he was not yet 50—had developed neurotic fears of his own death. Worrying that the still wobbly edifice of Freudian psychoanalysis would topple if his fears were to materialize, he felt the need to designate and train a successor.

Because he and many of his early adherents were Jewish, he was concerned that psychoanalysis would be tarred as a Jewish movement; so he deliberately chose a non-Jew as his potential successor. This was Carl Gustav Jung, a boisterous, brilliant Swiss psychiatrist with a strict Calvinist background and mystical leanings. Psychoanalysis thereby began its slow movement out of the confining limits of the Vienna circle; Zurich was its first stop.

A colleague of Freud's, seeing elements of Jung's personality and beliefs that presaged trouble, had warned him early on that Jung was a poor choice. But Freud, unheeding, not only hung on but poured great barrels of love into the relationship. At one point, already chafing under Freud's leash, Jung quoted Nietzsche: "One repays a teacher badly if one remains only a pupil."

Something, however, had bound the two men together in a relationship that was deeply personal as well as professional, to the point that they took on the roles of "father" and "son," though in a mixed-up way. Freud was the undisputed leader of the small circle that used to meet in his quarters every Wednesday night—he was the father of whom all the "children" wanted to be the favorite.

But with Jung something was different, and they exchanged roles. In one case, Freud fell over in a dead faint during a discussion of this relationship, and Jung carried him to a couch. Later Jung recalled that "in his weakness he looked up at me as if I were his father."

As with many family relationships, this one had elements of both love and hate. Of this and another such fainting episode in the presence of Jung, Freud suggested that he hated Jung because of what he perceived as Jung's death wish toward him. Jung, for his part, commented: "Just like a woman. Confront her with a disagreeable truth: she faints."

Jung had his own ideas on psychiatry, however, and began to question Freud's emphasis on sexuality as a basic cause of neurosis; he was sure it was "inadmissible biologically." [30] He also argued for dogma and ritual as requirements for psychological health. Freud, a combative atheist, was repelled by such ideas.

So, when Jung—brilliant and ambitious—chose to move out on his own, it was more than Freud could take. But the parting, due at least partly to personal conflict, was hard for both of them.

Slowly and painfully, respect and love turned to resentment and derision, and the deep, loving relationship turned into a bitter enmity that lasted long past their eventual breakup.

Neither was free of blame. Even as Jung was putting forth his own ideas he claimed, at first, that he was not deviating from orthodox psychoanalysis. Later, he became, ironically, the major spokesman for psychoanalysis after Freud himself, and, thanks mainly to Freud, had been installed as president of the International Psychoanalytic Association. Yet Freud feared that Jung and his followers might even force him and his adherents out of the association, and he began a campaign to counter this possibility.

The Freud-Jung story involves a fascinating amalgam of factors and influences: the acknowledged neuroses of the two men and their often wild analyses of each other's activities and dreams; the machinations of brilliant men who managed to inflate even accidental slights and inferences into major insults; the pushing and pulling of their associates, including a series of other feuds between the two men and others in the movement; the national pride of and competition between Zurich and Vienna; the differences between their beliefs (including Freud's reactions to Jung's incorporation of astrology and the occult in his work, and then Freud's growing interest in the psychology of religion); and the personal demons that both drove them together and then drove them apart.

At the same time we are delighted by the occasional insightful recognition of potential problems. In a 1911 letter to Freud, Jung acknowledged that it's "a risky business for the egg to try to be cleverer than the hen." [31]

Other fascinating implications and interplay: As Freud became more involved with the psychology of religion he began to investigate the story of Amenhotep (and other Egyptian pharaohs) who erased their fathers' names from inscriptions and substituted their own. Freud compared this with what was happening in their relationship, suggesting that Jung was trying to wipe out Freud's name and replace it with his own.

By 1913, though they maintained a cover of courtesy, the goings-on backstage were another matter. Freud complained that Jung was "outrageously insolent," and had revealed himself as the "florid fool and brutal fellow that he is."

The hurts continued well past the break, which took place in episodes that stretched from about 1912 to 1914. Freud later wrote of Jung's "lies, brutality and anti-Semitic condescension to me." In a 1924 letter he called Jung an "evil fellow." And as late as the 1930s, Freud would still mutter, "Bad character," at the mention of Jung's name.[32] To the end, he dismissed Jung's work.

For his part, Jung later complained that Freud was unquestionably a neurotic and he talked about Freud's sourness and anger. But his view of Freud continued to waver between deep respect and strong resentment until his death in 1961.

Up, Up, Up and Down, Down, Down

Freud's movement started slowly. His *Interpretation of Dreams* sold only a few hundred copies. But in spite of all the opposition, and all the bickering, both the movement and his reputation took hold and grew steadily. By the middle of the 20th century his ideas had built into a mighty, unstoppable wave that far outstripped any criticism, especially in the United States.

But for Freud himself, the story is different. Although he somehow managed to squeeze into his extraordinary schedule some solace and pleasure from his family life, his professional life was a constant round of struggle and conflict. Regardless of what Freud's life seems to us to have been in the first decade of the 20th century, Jones calls them

> his last happy years. They were immediately followed by four years of painful dissensions with the colleagues nearest to him; then by the misery, anxiety and privation of the [First World] war years, followed by the total collapse of the Austrian currency with the loss of all his savings and insurance; and, very little later, by the onset of his torturing illness (cancer of the jaw) which finally, after sixteen years of suffering, killed him.[33]

Add to this the years leading up to World War II, years of increasing violence, of continued and ever-growing anti-Semitism. Though his international reputation helped spare his life, he was forced to uproot himself and his immediate family and, a tired and ailing man, flee to England; his four sisters later perished in the Nazi death camps.

In the same mid-century years, he faced yet another difficulty in his professional life. As psychoanalysis gained success and became an established method of treating mental illness, the medical world tried to monopolize it, insisting, especially in the United States, that only medically trained psychiatrists should be permitted to carry out psychoanalytic treatment of patients.

Freud then found himself in a strange position. On the one hand, he was insisting that lay analysis was an acceptable, even desirable, approach—that competent, successful analysis could be carried out by non–medically trained analysts. Yet at the same time he wanted desperately for psychoanalysis to be a true, if independent, science of the mind—quantifiable and testable—and perhaps in that way to gain acceptance from the medical establishment. He recognized that it was not there yet, but always hoped that this could be made to happen.

But the dichotomy remained a source of conflict, and continues on until today. Perhaps some latter-day Freud will come along and find a way to complete his dream.[34]

Freudians versus Fraudians

Still, his reputation seemed secure. Then, around the middle of the 20th century, new psychoactive drugs began to hit the market. Lithium, used to treat manic depression, appeared in 1949 and opened up the field; then other antipsychotic and antidepressant drugs, including the phenothiazines—such as Thorazine—and the tricyclics—such as Tofranil—appeared. The psychiatric community quickly saw a revival of the 19th-century approach, treating the brain rather than the mind—the revival of psychiatry as neuroscience.

As with any new development, the reactions were extreme: talk therapy is baloney; mental problems are chemical or biochemical imbalances of some sort and, with further development of drug therapy, mental illnesses will all go away; Freud's method may have been useful in the predrug era but is superfluous now.

Of course, it hasn't worked out that way. At the very least, or so far anyway, drugs have proven themselves useful mainly in the more serious disorders, generally classed under the term "psychosis," with bipolar disorder and schizophrenia being the main ones. Nevertheless, pills were prescribed for an ever-widening range of mental and emotional problems, and by an ever-increasing percentage of the general medical establishment. Miltown, Librium, Valium, Xanax, and Prozac became as familiar as aspirin.

The field of psychiatry found itself under attack from many quarters—by the medical establishment on one side, and by lay analysts on the other; by a backlash against the simpleminded, timesaving, take-a-pill approach; and by writers such as Thomas Szasz in the United States and Michel Foucault in France, who even argued that there was no such thing as mental illness in the first place.[35]

Medical students began to pull back from the field. In 1984, 3.5 percent of American medical graduates planned to specialize in psychiatry; a decade later the figure had dropped to 2 percent.[36]

Psychoanalysis suffered, too. As Freud's dominance waned, criticism grew ever louder; and as the decades have passed, revisionist ideas have bloomed. Suddenly questions that had long lain dormant erupted. By the 1970s questioning of Freud's work and of Freud himself was common: Frank Cioffi began arguing that Freud's theories are inherently untestable, and therefore that psychoanalysis can never be more than a "pseudoscience."[37]

Freud himself began to take a drubbing as well; revisionist biographies, such as those by Paul Roazen (1975) and Frank Sulloway (1979), emerged that contained descriptions of a very blemished human being, rather than the Olympian god that Jones and others had worshiped.

In an interview a Freud defender, Kurt Eissler, said sadly:

> No one is interested if you write positively about Freud, but as soon as you write negatively everybody applauds. . . . Imagine! Swales "proves" that Freud seriously planned to kill Fliess! . . . Roazen writes that Freud was responsible for Tausk's suicide.[38] Krüll writes that Freud watched his father masturbate and suppressed the seduction theory in order to protect him. Masson writes that Freud dropped the seduction theory out of cowardice and to further his career.[39]

A Blemished God

Was Freud a shy, retiring gentleman who never looked for a fight, as Jones suggests in his biography?[40] Or was he more like Roazen's description of him:

> Freud's system of thought reflects his fighting stance; he used military language and the imagery of warfare throughout—attack, defense, struggle, enemy, resistance, supplies, triumph, conquest, fight. It was somewhat flat-footed of Jones to write: "I do not think he took the opposition greatly to heart." In some sense, he obviously quested for the antagonism he stirred up. . . . [41]

Another source of conflict—highly significant or strongly questionable, depending on your point of view—heaves into view. Born into a nonobservant but otherwise strongly Jewish family (his mother spoke Yiddish, not German), Freud carried on this tradition. Roazen says that among Freud's heroes were Moses and the biblical Joseph. No surprise there, but Roazen also mentions Napoléon, Hannibal, Cromwell, and Alexander the Great—in all of whom he sees a Jewish connection; for example, he describes Hannibal as a Semite rather than a Carthaginian.

Was Freud the victim of a virulent anti-Semitism? Sander L. Gilman, professor of Germanic studies at the University of Chicago, suggests that in both the science and the medicine of Freud's time, Jewishness was considered a kind of biological flaw, and that Freud's psychoanalysis emerged as his way of dealing with the biological discourse that was sloshing around him.[42]

Ernest Jones argues that Freud lived "in a city ruled by the anti-Semitic Bürgermeister Lueger, and where anti-Semitism prevailed in professional, academic and governmental circles. . . ."[43]

According to Freud, it was just this experience that gave him the strength to overcome powerful opposition. He says that when he entered the University of Vienna in 1873, "I found that I was expected to feel myself inferior and an alien because I was a Jew. . . . These first impressions at the University . . . had one consequence which was afterwards to prove important; for at an early age I was made familiar with the fact of being in the Opposition. . . . The foundations were thus laid for a certain degree of independence of judgment."[44]

Freud also claimed, in a letter to his friend and colleague Karl Abraham, "You may be sure that if my name were Oberhuber my new ideas would despite all the other factors have met with far less resistance. . . ."[45]

Not so, say others. In fact, Jones himself is not sure; he compares the response to Freud's ideas with "my own experience in England where we found quite enough 'resistance' although in the first dozen years there were only two Jews in our Society."[46]

Roazen argues, more strongly, that Freud was "extraordinarily sensitive to criticism" and tended to "exaggerate the extent and intensity of opposition to him and his work, just as he probably made too much of the role anti-Semitism played in his life."[47]

Roazen adds: "To him [Freud], opposition was a sign of recognition. It is difficult to ascertain which came first, Freud's cosmic provocativeness or the savage attacks on him."[48]

As is often the case with complex people, there is an intermediate approach. Hanns Sachs, a contemporary of Freud, wrote that in the execution of his duty, Freud was "hard and sharp like steel, a 'good hater' close to the limit of vindictiveness." Yet Sachs could also report of his own years with Freud: "I never heard him raise his voice in anger or excitement."[49]

Another question: Was Freud rejected by orthodox medicine and psychiatry right from the beginning, as the traditional story has it, or did he distance himself from it? Of the mid-1880s, when he set himself up as a specialist in nervous diseases, he later wrote: "As I was soon . . . excluded from the laboratory of cerebral anatomy [where he had done some work] and for a whole session had nowhere to deliver my lectures, I withdrew from academic life and ceased to attend the learned societies."[50]

Finding his professional ostracism depressing, he sought other company. In 1895 he joined the B'nai B'rith Society, a strongly Jewish-oriented social organization, in which he maintained membership for the rest of his life.[51]

Sulloway, however, argues that Freud really wasn't all that isolated professionally, that he was in constant touch with a variety of colleagues. Sulloway adds, "It is only the later rise of the psychoanalytic movement (together with Freud's destruction of most of his pre-1907 correspondence) that has made this early period seem like a state of heroic isolation."[52]

On the other hand, it seems hard to argue with the following, which recalls Harvey's experience after he came out with his surprising ideas. Freud, in a letter of May 4, 1896, to Wilhelm Fliess, complained: "Word was given out to abandon me, for a void is forming all around me. So far I bear it with equanimity. I find it more troublesome that this year for the first time my consulting room is empty, that for weeks on end I see no new faces, cannot begin any new treatments, and that none of the old ones are completed. Things are so difficult. . . ."[53]

More Questions

The traditional story is that Freud unleashed infantile sexuality on a prudish, repressed society. Jones writes of "the tired and narrow-minded atmosphere of Vienna. . . ."[54] Freud complained: "Few of the findings of psychoanalysis have met with such universal contradiction or have aroused such an outburst of indignation as the assertion that the sexual function starts at the beginnings of life and reveals its presence by important signs even in childhood. And yet no other findings of analysis can be demonstrated so easily and so completely."[55]

Oliver Sacks, the well-known neurologist and writer on neurological oddities, agrees. When asked whether Freud's insistence on the centrality of sex and the sex drive was one of the reasons why his ideas were resisted so strongly, Sacks answered, "Yes. It seemed outrageous, and when Freud threw in infantile sexuality as well that seemed even more outrageous. . . ."[56]

Finally, Jones states: "In those days Freud and his followers were regarded not only as sexual perverts but as either obsessional or paranoic psychopaths as well, and the combination was felt to be a real danger to the community. . . . No less than civilization itself was at stake. . . ."[57]

Yet here again there are alternative views. Bruno Bettelheim, himself a Freudian, writes: "In this unique Viennese culture, the strongest inner powers were thanatos and eros, death and sex."[58] He refers, for example, to the suggestive and often tortured works of the artists Klimt, Schiele, and Kokoschka, as well as various other works in art and music. Freud, in other words, was not operating in a sexually repressed atmosphere. (Freudians would counter that Schiele was imprisoned for a brief period in 1912 for obscenity.) Sulloway (coming down on Moll's side) also argues that Freud was not even the discoverer of infant sexuality.

As for the idea that Freud was the "discoverer" of the unconscious, Sulloway approvingly quotes from British philosopher Lancelot Law Whyte, who wrote in 1960 that "the general conception of unconscious mental processes was *conceivable* . . . around 1700, *topical* around 1800, and *fashionable* around 1870–1880."[59]

Among the major claims against Freud is that he used his own experiences (e.g., his own self-analysis) to make broad generalizations. Although Freud claimed in his *Three Essays on the Theory of Sexuality* that he had "carefully avoided introducing any preconceptions," Sulloway argues that that is exactly what he did not avoid.[60]

But these were only warm-ups for the fireworks yet to come.

Freud Bashing Is Fun

By the 1990s, Freud bashing had become an industry, with book after book and article after article going after him and his legacy with the vituperation usually relegated to child molesters. Author Richard Ofshe recently described Freud as "a Viennese quack distinguished only by a certain low cunning and a cigar."[61]

Frederick C. Crews, a leading Freud basher, suggests in his book *Unauthorized Freud: Doubters Confront a Legend* (1998) that "our great detective of the unconscious was incompetent from the outset—no more astute, really, than Peter Sellers's bumbling Inspector Clouseau—and that he made matters steadily worse as he tried to repair one theoretical absurdity with another."[62]

Nor are such feelings the rantings of a few lone doubters. Members of whole groups, such as the feminists, have been heard from, arguing against a variety of Freud's ideas.[63] The respected *New York Review of Books* has joined in the media madness and has published several major anti-Freud essays, which in turn have generated a veritable frenzy of pro and con responses.[64]

In 1996 a major exhibition on Freud's life and work was being mounted at the Library of Congress; it was brought to a screeching halt by a petition signed by 50 major anti-Freudians—including Freud's own granddaughter—all of whom feared it presented Freud in too positive a light. The exhibit did finally open two years later, but with some significant changes.

By now, everything is up for grabs. Even the concept of Freud as a tidal wave is questioned. Professor Adolph Grünbaum of the University of Pittsburgh challenges the premise "that Freudian theory has become part of the intellectual ethos and folklore of Western culture . . . the prevalence of vulgarized pseudo-Freudian concepts makes it difficult to determine reliably the extent to which genuine psychoanalytic hypotheses have actually become influential in our culture at large." The example he uses is that of Freudian slips. These, he argues, are commonly applied to *any* bungled action, even though Freud actually had in mind only those whose motives are unknown to consciousness.[65]

I have not felt it appropriate, or necessary, to provide answers to the various challenges to Freud and his work. There are plenty of Freud supporters who happily provide answers to all of them. One of the best general defenses I've seen is a long, reasoned article by Jonathan Lear of the University of Chicago, who concludes with this idea:

> Beneath the continued attacks upon him, ironically, lies an unwillingness to let him go. It is Freud who taught that only after we accept the actual death of an important person in our lives can we begin to mourn. Only then can he or she take on full symbolic life for us. Obsessing about Freud *the man* is a way of keeping Freud *the meaning* at bay. Freud's meaning, I think, lies in the recognition that humans make more meaning than they grasp, that this meaning can be painful and disruptive, but that humans need not be passive in the face of it. Freud began a process of dealing with unconscious meaning, and it is important not to get stuck on him, like some rigid symptom, either to idolize or to denigrate him. The many attacks on him, even upon psychoanalysis, refuse to recognize that Freud gave birth to a psychoanalytic movement which in myriad ways has moved beyond him.[66]

Peter Gay adds: "The very fierceness and persistence of his detractors are a wry tribute to the staying power of Freud's ideas."[67]

Still, there remains the unanswered question of whether Freud was really given a hard time or whether he imagined this to be the case and then gave it this spin in later years. Norman Kiell, who did an extensive study of the early reviews of Freud's work, makes the point, first, that "the early reviews . . . reveal how inept they were. Neither the psychiatric nor the lay reviewers seem to grasp what Freud meant or intended. . . ."[68]

Only then does he put forth his basic conclusion: namely, that "the first reviews by the psychiatrists were totally negative. . . ."[69] He also refers to "the frequent target [Freud] became for colleagues . . ."[70] and adds, "While the lay reviewers very rarely flayed Freud, it was the exegetical professionals and amateurs in their polemical articles and books who vituperatively savaged Freud, as did some of the early psychiatrists in their reviews and congresses."[71]

This clearly is not the last word on the question, but it does bolster Freud's claim.

Perhaps we've reached a kind of standoff in which the argument becomes repetitive as contemporary challengers think up attacks that are really just rehashes of earlier ideas that have become buried in a mass of Freudiana so vast that no one can be aware of it all.

While researching this chapter, I called and asked a young library intern to run through my account and see which books I could renew. After hearing her mispronounce Freud's name as Fried a couple of times, I gently suggested that his name was pronounced "froid."

Her next attempt came out as *Fraud.* I figured that was close enough.

Fried and Fraud. Freud would have had fun with that one.

CHAPTER 8

Sabin versus Salk

The Polio Vaccine

The smallpox vaccine, developed by Edward Jenner at the end of the 1700s, is not named for that scientist. Similarly, the yellow-fever vaccine, developed shortly before World War II, is not named after Max Theiler, its developer.

But the first vaccine for poliomyelitis came quickly to be called the Salk vaccine, and Jonas Salk himself became a national icon of heroic proportions.

Yet Salk knew that this meant trouble, that his colleagues in the biomedical sciences would not be happy about it, and that he would be attacked as a "glory hound." Which is just what happened.

One problem was his strong desire and drive for professional recognition, which was evident in everything he did. This seemed to annoy many of his peers—who were no less guilty of the same behavior. Still, the publicity was thrust upon him. In fact, he tried to back away from the public adoration and even refused at times to cooperate with the press and with others who were trying to make him a national hero. This led to additional damage to his reputation, which would last for years.

In his own words: "The worst tragedy that could have befallen me was my success. I knew right away that I was through—cast out."[1] Acclaimed by the public as the new Pasteur, he was at the same time practically excommunicated by his own peers.

Among the most vociferous critics was Albert Sabin, a well-known medical scientist who was developing a vaccine of his own. Richard Carter, one of Salk's early biographers, writes: "The professional controversies that obstructed the development and testing of his [Salk's] killed-virus vaccine were waged with the intensity that man usually reserves for his holy wars. The brilliant, articulate Dr. Albert Sabin attacked Salk's work . . . on the front page of every important newspaper in the U.S.—not once but often."[2] Among Sabin's pithier comments: "You could go into the kitchen and do what he did,"[3] and "he never had an original idea in his life."[4]

The treatment of Salk by his colleagues leads us to an initial inclination to leap into his corner, as exemplified by the Carter quote in the previous paragraph. Though the story sounds like a simple one at first, involving little more than sour grapes on the opposing side, we will see that there was far more to the story than this.

A Puzzling Disease

Though there had been earlier reports of polio, multiple paralytic cases began to break out here and there toward the mid-1800s: several cases in England in 1835, Louisiana in 1841,[5] and the island of St. Helena in 1844,[6] and then, more widely, in Scandinavia in the 1880s and 1890s.[7] Over the next half century, polio epidemics scarred many nations worldwide—including, perhaps especially, the United States.

Although poliomyelitis never killed with the ferocity of such epidemic diseases as bubonic plague, cholera, typhus, and typhoid,[8] it provided a unique terror of its own. With other major diseases, those who die leave behind grief and anger, but they themselves are gone. With polio, vast numbers remained alive but severely handicapped, taxing the caregiving capacities of both families and the medical world. Worst of all, most victims were children, leading to the nickname infantile paralysis. Anyone over 50 can vividly remember youngsters in ungainly metal leg braces, or totally immobilized and engulfed in iron lungs, needed just to keep them breathing.

No one knew how the disease entered the body and how it did its damage, and there seemed no way to stop its spread. Parents locked their children into their rooms, and even boarded up the windows to, hopefully, protect them from infected air.

Medical people in the Western countries were particularly upset because the disease was becoming a paralytic plague in the very same years that germ theory and scientific medicine were coming into their own; that is, when it was thought that the work of Pasteur, Koch, and others was transforming the world of medicine.

By 1908 researchers had shown that polio was caused by a virus and that it attacked the nervous system, though it was quickly also found in other tissues. Looking back at Pasteur's success with vaccines for several diseases, researchers thought it shouldn't be that much more difficult to find one for polio.

Yet while other viral diseases either waned or at least held constant, poliomyelitis was on the march. It was recognized as a major health problem by 1910 and then, in 1916, the incidence rate suddenly quadrupled. New York City suffered an onslaught of 9,000 cases, resulting in 2,343 deaths. The rate eased somewhat in subsequent years, but the panicky fear remained.

The disease—frightening, apparently unstoppable—brought out some of the worst in people. Neighbors blamed neighbors; hospitals refused patients for fear of further spread; treatment of black patients in the South was often separate and not always equal; suburbs blamed the cities.

Then, no less distressing, fingers began to point at recently arrived immigrants from southern and eastern Europe, assuming that they had brought the disease from their slums to the cleaner, better-off, middle-class neighborhoods of the United States. The slums of South Philadelphia and the Lower East Side in Manhattan were particularly suspect. Quarantine notices were tacked up outside tenements and multifamily dwellings, but not on private houses.

Children from slum areas who, due to lack of room, could not be isolated were often taken forcibly to hospitals, even though the parents feared they would be worse off there, where little could be done, anyway. People fled the cities, a time-honored technique, to no avail.

Here and there a health officer or an epidemiologist might note that the congested and presumably infected districts had no higher rates of the disease than the better-off districts elsewhere, and in some cases lower rates. But such ideas were swept under the rug in the hysteria and total inability to come up with anything solid. Health officials closed playgrounds and children's rooms in libraries; they disinfected sandpiles. Nothing seemed to work.

Had they dug a little deeper, they might have come up with a most ironic twist. Before about the 1890s most young children worldwide were exposed to a kind of nonparalytic polio and developed antibodies to it that protected them against the more vicious manifestations.

Then, as developed countries improved their sanitary conditions, children in these countries were no longer being exposed to any polio virus, and did not develop antibodies to any form of the disease. These kids were then easy targets for the paralytic form of the virus. Such diseases as cholera and typhus took a nosedive, but the poliomyelitis virus—which propagates somewhat differently, though this fact was not at first known—thrived and increased in potency.

In other words, children from clean, sanitary, well-ordered homes were at *greater* risk of the paralytic form than those from poorer, and dirtier, sections. But what self-respecting, patriotic, well-scrubbed health officer or physician was going to believe that? Some of those seeking a cause began to believe there must be some moral weaknesses in the victims that led them to be so afflicted.[9]

Competing Approaches

Virologists had little doubt that a vaccine was the only real hope. But two early attempts in 1935 led to disaster. In one case, a respected bacteriologist tried to apply the same techniques that Pasteur had used successfully some half a century before with fowl cholera, anthrax, and rabies. He developed a weakened-virus polio vaccine.

At about the same time another attempt was made using a killed virus. The basic method here is to purify and culture a virus, then chemically inactivate it in such a way that it no longer can cause disease, yet retains the ability to generate an immune response.

Both methods seemed at first so promising, and the need so great, that they were tried without proper preliminary testing. Not only did they not help, but there were cases resulting directly from the vaccines, including paralysis and even several fatalities. The only positive outcome was the lesson that this was a complicated affair and needed more research.

Salk and Sabin later became the standard-bearers for these two major competing approaches to developing a polio vaccine. Salk championed the killed-virus method, Sabin the weakened virus. Each had good reasons for his approach.

In the mid-1930s, Salk, while still in medical school, heard the standard pitch that immunization against viral disease calls for vaccines made from living, but weakened, infectious viruses. As in the earlier experience with smallpox, the innoculation causes an actual, but mild, infection, thereby stimulating the immune system to build up its defenses in advance of any later serious infection.

But he also learned that vaccination with killed organisms can work with bacterial disease; that, for example, the lethal diphtheria toxin could be made into a killed, noninfectious vaccine by treatment with a form of a standard disinfectant and preservative, formaldehyde.

The idea of using this method for viral disease lodged in the back of his mind and led to his later work with killed viruses on polio. Along the way, he did some useful, and honored, work with flu vaccines, using this approach. He argued, along with a few others, that the immune system could be set to work against polio viruses, too, without need to trigger an actual infection.

Sabin, like Pasteur, believed the best way to produce immunity was to induce a mild infection with a live virus, but one that has been weakened chemically or biologically. By the late 1930s it had finally been learned that the polio virus enters the system via the mouth and the digestive tract. This suggested that a satisfactory vaccine could be developed that could be taken by mouth, a far simpler and more expedient route than had to be taken with the killed virus, which required a shot. This method also mimicked more closely the natural route of infection.

Most virologists agreed with him, and also believed that the immunity conferred by a killed virus would only be temporary, at least with the polio virus. (And as it turned out, each child did have to take several shots plus a booster for a complete treatment with Salk's killed virus.)

Finally, the more "natural" infection caused by the weakened-virus vaccine can spread, but this time in a good sense. Because the polio virus enters the body via the digestive tract, it is shed in stools and can then spread either by direct contact or through swimming pools or water systems. In most cases of infectious disease, this is a major route of infection. With the weakened-polio virus, however, the "infection" transferred from an inoculated child will then *protect* the "catchee." This does not happen with the killed virus.

These points convinced Sabin, and most of the research establishment, that the weakened virus was the way to go, despite the fact that a killed-virus type could be developed more quickly.

Obstacles

The development process for both vaccine methods, however, was held back by several major obstacles.

An important one had more to do with political maneuvering than medical science. A few important groups held the reins and virtually controlled the direction of research. In the early and mid-1930s, Simon Flexner, an eminent medical researcher at the Rockefeller Institute, led a group of scientists who were convinced that the polio virus grew only in the nerve cells of the spinal cord and brain. They were further convinced that it invaded victims via the respiratory system, as was the case with other viral diseases, including influenza and pneumonia. The group simply ignored inputs from both Scandinavian and American researchers who were closer to the clinical side and saw more clearly that the virus could be recovered from the intestinal tract and its contents. It was Sabin himself who, in 1939, provided convincing evidence that the virus entered via the mouth and digestive tract.

By this time researchers had also found that more than one polio virus could cause disease. The different forms were called strains. There appeared in fact to be many strains. (With both smallpox and yellow fever, only a single strain of the offending virus was involved.) This added to the complications. These strains, however, could hopefully be grouped into just a few types. Any successful vaccine would have to utilize all the types.

A committee was formed to deal with the problem and to set up procedures for typing, which meant figuring out which strains belonged to which type, and how many types there were. Both Sabin and Salk were on it, but already there was conflict in the air. The committee, Salk began to feel, was going off in the wrong direction. He believed that the typing process would end up far more complicated than it had to be because they were trying to set it up based on a strain's virulence he felt it could be done on the basis of its antigenicity—how much antibody resulted from introduction of the virus (as vaccine) or its infective portion, called the antigen.

The standard typing process, championed by the virology community (including Sabin), involved infecting monkeys with a known strain, then inoculating other monkeys with virus obtained from the first ones. The process required taking careful note of which survived, and with what results, in a complicated round of injections, and careful checks and balances. It was a dirty, complex, and costly business.[10]

At one of the meetings of the typing committee, in mid-1948, Salk tried to bring up his ideas, which leaned more toward immunology than virology. He later wrote:

> I was suggesting that our purposes might be better served by testing an unknown virus's capacity to immunize, rather than worrying about its capacity to infect. Albert Sabin sat back and turned to me and said, "Now, Dr. Salk, you should know better than to ask a question like that." It was like being kicked in the teeth. I had offered an oblique challenge to one of the assumptions, you see,

and now I was being put in my place. I could *feel* the resistance and the hostility and the disapproval. I never attended a single one of those meetings afterward without that same feeling.[11]

A year later new findings at Johns Hopkins and Yale established the number of polio virus types at three. But the findings had to be confirmed by the committee—a gargantuan task that involved the typing of 195 strains from near and far.[12]

In the process, Salk gained a greater familiarity with the polio virus and, with funds obtained via the typing program, he was able to equip a new laboratory at the University of Pittsburgh. He later recalled:

The typing program was to take three years, but our laboratory had the whole thing solved before the end of the first year. Everything that happened during the last two years was confirmatory. What could I do? I couldn't slap these people in the face and call them dumb bunnies and shriek that they were doing their job asswise. Even if I could have, I would not have wanted to. They had their way of looking at things and I had mine. . . . [13]

The Pittsburgh lab also provided Salk with a useful locale for his later work on the vaccine.[14] To be useful it would have to contain all three types.

Another obstacle had shown up even earlier, however. Viruses cannot be cultured in nutrient substances, as bacteria can. Sabin, in fact, had made a fairly serious misstep in 1935 when he and a coworker, Peter Olitsky (both then at the Rockefeller Institute for Medical Research), declared that their experiments showed clearly that the virus would only grow in living neural tissue. Though they had been very careful, they were using a strain of polio that had been passed through the nerve tissues of animals so many times (a way of weakening the virus) that it no longer "knew" how to live in other mediums.

As a result, the virus was for years only grown in nerve tissue from expensive laboratory primates; further, nerve cells from any creature are difficult to cultivate and manipulate in culture. All of which meant the virus was difficult to obtain and so set strict limits on research.

A Break

But time and science march on, and by 1948, John Enders, an American virologist at Harvard University Medical School, and two colleagues found a way to grow polio virus in human non-nerve tissue. They managed to culture the virus on tissue scraps obtained from still-born human embryos, and then on other tissue.

The result overturned the Sabin-Olitsky pronouncement. No longer were expensive laboratory primates needed for the research; a much more copious supply of polio virus became available for study and for preparation of vaccines. Researchers later found that when the virus was grown in this way it lost its deadly virulence, yet was still useful for vaccines.[15]

This was an important finding and Enders, along with his students and coworkers Thomas Weller and Frederick Robbins, was awarded the Nobel Prize in physiology or medicine in 1954. In spite of the enormity of the polio problem and the world's great sigh of relief when a successful vaccine was finally developed, it was to be the only Nobel Prize awarded for polio research.[16]

Salk was one of the first to recognize the importance of the Enders discovery, and he made adjustments in his lab's techniques to take full advantage. As he put it in a television interview: "Enders threw a long forward pass and we happened to be at a place where the ball could be caught."[17] He continued to work with killed virus.

Sabin, who actually had more virological experience than Salk, had developed vaccines against dengue fever and Japanese encephalitis; he began working on attenuated strains of polio virus in 1953. Despite the tragic results of the 1935 debacle with a weakened vaccine, this was still the favored method of the research establishment. Enders sided with Sabin on this issue, and in fact is reported to have referred to Salk's work as quackery.[18]

So the stage was set for a major competition. By 1952 researchers had already shown that either kind of vaccine caused a rise in blood antibody levels against the virus. But no one knew yet what level was needed to provide an effective block. Salk's approach, a kind of all-or-nothing attack on the virus, was quicker, faster, and, by roughly 1952–1953, just about ready to go. Sabin's (and a few others') weakened type still needed considerable fine-tuning.

The National Foundation for Infantile Paralysis

What does it take for a major development like a polio vaccine to come to fruition? Is it a matter of a lone, brilliant scientist going into his or her lab a couple of times, making the discovery, and emerging a national hero the next day? Rarely. We've already seen that the scientific road to the polio vaccine was partially paved before Salk entered upon it.

When dealing with the health field, however, a new development is not enough in itself; it must be tested, and on a large scale, before it is made widely available to the public. Participation can come from unlikely sources.

In 1921, five years after the United States suffered through a major epidemic, a budding politician named Franklin Delano Roosevelt contracted polio at the age of 39. His brave battle with the disease, and, especially, his ability to go on to a sensational political career even though paralyzed from the waist down, are worthy of all the press it received. Some of his admirers feel, in fact, that his special abilities were honed in his battle with this major handicap (not that this wasn't covered up at every turn).[19]

But his affliction also got him interested in the disease. In 1926 he established a polio treatment center in Warm Springs, Georgia, and for a while he even participated in running it.

By 1928 he decided it was time to embark on the political road and turned to his close friend and law partner, Basil O'Connor. O'Connor's first job was to obtain funds for keeping the expensive center solvent. Then, as time passed, he moved on to the far bigger job of helping pay for both treatment of those already ill, and research toward prevention of this monstrous disease, throughout the country.

Among the innovative moves initiated by O'Connor and other friends of Roosevelt was creation of the National Foundation for Infantile Paralysis, set up to coordinate and help finance the whole polio effort. It proved to be a good model for other large-scale voluntary health groups, such as the American Cancer Society and the American Diabetes Association. The foundation sponsored and ran the astonishing grassroots movement called the March of Dimes, probably the one time other than in war years that virtually the entire country pulled together.

The foundation raised large sums during the decades of its existence, though much of its effort had to go toward helping in the massive costs of dealing with outbreaks, and toward development of new and better treatment modalities. To the credit of both O'Connor and the foundation, the urgent need for *prevention* of the Crippler, as it was often called, remained an important goal.

By 1950 the foundation decided the time had come to put its research eggs into a vaccine basket. But which one? Neither side had anything good to say about the other. Nor was all the complaining one-sided. Sabin later complained that Salk's "behavior on a number of occasions has been less than suitable for friendly relations." [20]

A scientist present at the time of one of the many polio meetings later recalled that Sabin and Salk just "seemed to rub each other the wrong way." [21] O'Connor was well aware of the antipathy between the two sides; fortunately, the foundation supported research efforts by both Salk and Sabin.

There the matter lay—until a fateful meeting between O'Connor and Jonas Salk aboard the ocean liner *Queen Mary* in 1951. Both were returning from a major international polio congress in Copenhagen. Salk, considered a kind of interloper in the virological world, had been treated at the congress with condescension, and he undoubtedly saw in O'Connor a way to break out of this squeeze. They had met before at conferences, but the leisurely trip seemed to make a difference. O'Connor was impressed by Salk's knowledge and, by then, experience in the field.

At the same time, while modern medicine was moving along nicely in many other areas, polio epidemics continued their deadly march. O'Connor's own daughter had been struck not long before and was partially paralyzed. Listening to and warming to Salk, O'Connor became convinced that a crash program, using a killed-virus vaccine, was both needed and a realistic possibility. Two main factors in its favor: it was virtually ready to go, and this type of vaccine could not cause the disease.

Salk could now proceed with actually preparing a vaccine that would be tried out. By spring 1952 he requested permission to test his vaccine first with polio victims at the D. T. Watson Home for Crippled Children, near Pittsburgh, to see how well the vaccine jacked up the antibodies in the particular strain they suffered from. Then he tested it on a group of inmates at the Polk State School, on children who had not had polio. Tests taken at the end of the year showed that antibodies in both groups of children had remained high, and there were no untoward effects in either group.

The National Foundation decided to organize mass trials using the killed-virus vaccine. The more conservative members of the polio community were horrified and angered to the point that Sabin and others began to publicly challenge the killed-virus approach, and even went after Salk himself.

The antipathy between the two men is somewhat ironic, for they had much in common. Both, at least early on, were considered intruders in the lofty world of medical research; both, being Jews of eastern European descent, were considered "ambitious."[22] Too, they both went to New York University Medical School. Later, both simply handed over their polio vaccines to the world, gratis.

The similarities end there.

Field Trials

Even aside from all the internecine warfare, the gearing up for the trials did not go smoothly. There were serious disagreements over who should have control; and there were problems with scaling up from laboratory production to industrial quantities. In fact, when tests were run on the early production quantities at the National Institutes of Health in March 1954, five of the first six batches were found to cause polio in the monkeys on which they were tested.

Paul de Kruif, the well-known author and bacteriologist, caused some unease when he leaked some of this to Walter Winchell, at the time an enormously influential columnist and radio personality. Winchell suggested that the vaccine could be a killer. And all the while Sabin and others were arguing against the whole approach.

But the numbers were getting worse each year: growing from about 35,000 cases in the United States in 1946 to a particularly vicious epidemic in 1952 when nearly 58,000 fell ill; 3,000 died and 21,000 were left paralyzed. The public clamor for a vaccine overrode any objections and the foundation decided to begin large-scale human field trials.

But this is a ticklish business, and the cooperation and sympathy of the medical profession was absolutely necessary. This cooperation was not a foregone conclusion, considering the objections that were still being raised[23]—and which rose in decibel level when the idea of a large-scale trial was presented at a meeting held in February 1953 at the Waldorf-Astoria Hotel in New York.

The governing group decided to submit an article to the prestigious *Journal of the American Medical Association* detailing the reasoning behind the killed-virus approach, along with progress made to date.[24] The idea was to make sure the medical community was informed before the general public was; this was to prevent embarrassment should a physician's patients learn of the vaccine before the physician did, and even demand inoculation before it was ready. If newspapers later summarized the information, that would be just fine, and would even help with fund-raising.

Unfortunately, Earl Wilson, a *society* columnist, got wind of the goings-on and broke the story. Everything the committee feared happened, and more. Salk was blamed for the leak; then another newspaper story suggested that Salk and/or the foundation had entered into some kind of deal with Parke Davis, a large pharmaceutical company, which was not true.[25]

He was seen as a glory hound; the more the newspapers wrote about him and the more the public adored him, the more his standing sank among his colleagues. He tried to call his preparation the Pittsburgh vaccine, but the newspapers thought, no doubt rightly, that "Salk vaccine" had a better ring to it.

The annual convention of the American Medical Association was held in June of that year, with 38,500 delegates and guests in attendance. Sabin addressed a major session and opened with these words: "Since there is an impression that a practicable vaccine for poliomyelitis is either at hand or immediately around the corner, it may be best to start this discussion with the statement that such a vaccine is not now at hand and that one can only guess as to what is around the corner."[26]

Then there were the inevitable objections and suggestions from the rest of the polio research world. The National Institutes of Health wanted to include an antiseptic in the vaccine; Salk argued that it would hurt its effectiveness.

The field trials began in the spring of 1954. At a major conference in Rome in September there was little of the collegiality one would expect at such a meeting. Aaron Klein, who wrote of the controversy in 1972, reported that Salk, Sabin, and others spoke in turn, "each defending his position and taking every opportunity to destroy the opposition. There was little of the compromise and mutual exchange that were supposed to occur at scientific meetings. Minds were closed: there was only attack and parry."[27]

Nevertheless, the trial began, an elaborate double-blind study involving nearly 2 million children. There were plenty of problems—including some bad batches from one of the manufacturers and, as a result, over 200 actual cases (a later count) that were in some way associated with the vaccinations; about 150 cases resulted in paralysis, 11 in death. Was there a problem with the specific batches produced by Cutter Laboratories, or with the vaccine itself?

It was a terrible time. Charges and countercharges flew. The fear and uncertainty were intense, at least until the problem was found to be limited to something—no one ever figured out what—that had happened to the Cutter batches. The unused portions were, of course, immediately pulled back and the Cutter production was stopped.[28]

But the locomotive had been started, and somehow the managing group, led by Thomas Francis Jr. of the Rockefeller Institute, pulled things together and restored some confidence in the program: 440,000 children were inoculated, 210,000 received a placebo, and 1,180,000 served as unvaccinated controls. Those trials remain the largest controlled, double-blind study in medical history.

A World-Class Event

By 1955 there was no question that the trials had been a success. A major, full-scale press conference was convened, and on April 12, 1955, the world was informed that a way had been found to end the disease's slaughter.

Speaking directly to an audience of 500 scientists and physicians, Francis presented the results in a careful, information-packed talk that lasted an hour and forty minutes. Sixteen television and newsreel cameras recorded the event, while in a press room three floors above, over 150 newspaper, television, and radio reporters were sending out details.

Salk was present and his name was much a part of the conference. He also gave a radio interview. But, several reports state, something was seriously awry. The previously sensitive, sharing, socially conscious scientist seemed now to be unwilling, perhaps unable, to share any credit for the great advance. Had the earlier adulation worked its wily ways? Was he hypnotized by the magic of the moment? Whatever the reason, in his remarks, then and later, he has been accused of making no mention of the work of John Enders, or of Isabel Morgan,[29] or of several other important predecessors and/or coworkers.

He was labeled selfish, a tag that stuck with him in the world of science. Medical historian Sherwin B. Nuland writes: "An egotism appeared that seemed not to have been evident before. The strongest charge leveled against the young superstar was his inability or unwillingness to share credit. . . ." Also: "None of those who had established the principles and the practices upon which the new vaccine was based was granted a share in the public acclaim."[30] Wilfrid Sheed, the well-known novelist and essayist (who fought his own battle with polio), comments on the press conference: "Salk took part . . . and went on radio but gave credit to nobody, including himself—of course, he was going to get the credit anyway. And that was the mistake that would haunt him." [31]

This take is a little troubling. The work of Enders, Morgan, and others had been well reported by then. The *New York Times*'s coverage of the press event, which spanned most of five full pages, included a full column on Enders's work, plus a short item on Salk's assistants at his own laboratory. [32] Though Salk's name was certainly prominent, only a few direct quotes from him were included.

Further, as with any press interviews, the interviewer has a strong part in what's covered, and Salk's omissions, both then and later, may not have been entirely his fault. Nuland himself adds: "What [Salk] was not able to overcome was that fact that his transformation from promising young virology researcher to the conquering hero of polio did not sit well with his collegues." [33] How much of his later reputation for immodesty came from this is hard to say.

To the press and the public, however, he was a national hero.

The Sabin Saga

Sabin continued to maintain that the Salk vaccine was actually harmful; he reasoned that while the immunity lasted, the body would not have a chance to build antibodies from natural exposure and that in the long run this would prove disastrous.

By 1956 he had a vaccine in hand and had tested it on monkeys and over 100 human volunteers. Results were exactly as desired: a significant rise in antibodies against the polio viruses and no signs of unwanted infection.

Sabin's opportunity to show his stuff came first with a successful trial in a seriously stricken Belgian Congo in 1958, and then in a massive trial shortly after in the Soviet Union. It is an indication of just how desperate the Soviets were that they decided to use a vaccine developed in the hated West.

On August 24, 1960, the surgeon general of the U.S. Army recommended the licensing of Sabin's attenuated vaccine for domestic use. In spite of this, and in spite of the successful large-scale trial in the Soviet Union, Salk continued to argue that that route was both "unnecessary and ill-advised," [34] that it could actually lead to some cases of polio from the vaccine itself.

He was right, but there were strong countervailing factors, especially in the less-developed countries. The Sabin vaccine was far easier and cheaper to use, a major factor in large parts of the world. Further, as I noted earlier, it produced a kind of immunity that was "catching." The killed vaccine, it was argued, was okay as a stopgap measure, but should now be set aside. So, even in the developed world, including the United States, the admitted advantages of the weakened-virus vaccine began to look better and better, and by 1961 the Sabin vaccine was the accepted standard around the world. Three decades later, health officials announced that the scourge had been eliminated from the North American continent.

During the years 1995–1996 some 400 million children under the age of five were immunized against polio, mostly with the Sabin vaccine. By 1996 the World Health Organization could report that there were less than 2,200 cases per year worldwide, the first time this could be said for over a century. Authorities now hope that they can eradicate polio worldwide, in this way duplicating the success achieved with smallpox two decades earlier. It won't be easy; in the poorer, hotter countries, there have been storage and administration problems. In India, for example, some batches of vaccine were ruined due to exposure to excess heat. But there is the hope that, as soon as the next few years, the Crippler can be beaten as a major health factor.

The more developed countries, such as the United States, face a somewhat different problem. Here, even a few cases stand out strongly. The objections voiced by Salk and others on his side have proved only too true. For reasons unknown, a small number of those who are given the Sabin vaccine continue to come down with the disease. Perhaps they are in some way genetically susceptible; no one knows why it happens. And although the vast majority of those who are immunized benefit, the small percentage who are struck can be struck hard.

Between 1969 and 1983, for example, the Centers for Disease Control (CDC) reported a total of 210 cases, 99 of which may have been caused by the Sabin vaccine. Sabin, however, refused to accept this. "There are other kinds of paralysis that simulate polio," he argued.[35]

Salk charged that "everyone—all but one man—believes the evidence. It is remarkable, simply remarkable, that he casts doubt upon it. . . ." Then Salk added: "What is surprising is that people accept it as the price we have to pay. I don't understand that, especially when it's clear that there is an alternative way to immunize which doesn't extract a price."[36]

These cases have made life difficult for groups, such as the American Academy of Pediatrics, that set up the inoculation standards to be followed in the clinical world. For a while, a course of inoculations was given that combined both types. Still, there were a few cases that showed up.

As a result, the U.S. government declared at the end of 1999 that the Sabin vaccine, regardless of its simplicity of administration and other admitted advantages, will be dropped in this country and replaced by a series of old-fashioned shots, using an upgraded—that is, stronger—version of the Salk vaccine.[37]

This, of course, would have pleased Salk—who died in 1995 at age 80—enormously. But we must remember that in the Sabin-Salk feud there is no real victor. Even though the Salk version provided both the opening and concluding scenes of this powerful drama, at least in the United States, it was clearly the Sabin version that has made this last apparent victory of Salk's vaccine possible.

Further, the oral vaccine will continue to be used domestically in special cases: to deal with any unexpected outbreaks; perhaps for children traveling to regions where the disease is still a problem; or in special cases for children who, for any reason, do not get the full series of shots.

Aftermath

Throughout the whole sad story of the feud, Sabin's reputation never wavered. In spite of all the nasty comments (on both sides), in spite of Sabin's fairly serious misstep in 1935, and in spite of the fact that he, too, never received a Nobel Prize, his reputation remained high in that fickle scientific world—as both medical statesman and researcher. After his vaccine was licensed, he went on to play an important role in the worldwide inoculation programs run by the Pan-American Union and the World Health Organization. He was researching the role of viruses in cancer at the time of his death in 1993, at age 86.

Unhappily, Salk's reputation never really recovered in the medical world. Though he received many honors from the public and from the government—for example, a Congressional Gold Medal in 1955 and the Presidential Medal of Freedom in 1977—there haven't been any major ones from the scientific community: no Nobel Prize, no election to the National Academy of Science. The usual explanation: he made no basic scientific discovery. Recall Sabin's caustic comment: "You could go into the kitchen and do what he did."

O'Connor, who was very close to Salk, complained, "He shows the world how to eliminate paralytic polio, and you'd think he had halitosis or had committed a felony."[38]

But Salk had no intention of drying up and blowing away. In a sense, he had to create his own medical world, which he did with the creation of the Salk Institute for Biological Studies in 1963. There, again with the aid of Basil O'Connor and the National Foundation, he was able to create a pure research institution where he and a cadre of research scientists could pursue their studies in an atmosphere of beauty and tranquillity.

It didn't quite work out the way Salk had hoped, however. In the decade it took to get the institute designed and into full swing, the research world may have pretty much passed him by. Jane S. Smith, in her fine retelling of the polio vaccine saga, writes: "When the dream institute was finally completed, . . . it became yet another source of misery and distraction. The great minds that Salk invited there turned out to have great egos to match; soon the host found himself dismissed as a 'mere technician' by the scientists he had gathered around him. . . ."[39]

Nevertheless, in the years since, the organization has done respected work in cancer, multiple sclerosis, and autoimmune disease, along with studies on the brain and the peripheral nervous system. At the time of Salk's death, he was working on a vaccine for AIDS.

Salk deserved better treatment. Nobel Prize–winner Renato Dulbecco wonders if Salk's experience could somehow be applied in the wider world of science. In an obituary for Salk, he wrote, "The fact that a fundamental advance in human health could not be recognized as a scientific contribution raises the question of the role of science in our society."[40]

Wouldn't a friendly competition have been better all around? The feud, in fact, may have been a factor in the Nobel Prize issue, particularly in light of a long-held suspicion by Nobel watchers that any hint of controversy is enough to scare off those who do the choosing.

On the other hand, suppose the competition had been friendly enough that one of them ended up convincing the other to switch sides?

Perhaps the feud was the better way after all, at least for the rest of us.

CHAPTER 9

Franklin versus Wilkins

The Structure of DNA

On April 25, 1953, a brief, 900-word article in the British journal *Nature* turned the biomedical world on its ear. Titled simply, "A Structure for Deoxyribose Nucleic Acid,"[1] it introduced the double helix to the world, and the authors became famous overnight.

For the first time in history nature's method of coding, and then passing on, all of the body's genetic secrets were suddenly opened for study on a molecular level. Note that this refers not just to human genes, but to every inherited characteristic of every living organism on earth—including disease-causing organisms!

And this meant that the medical world could begin to attack many of its challenges from below—challenges that include not only obvious hereditary diseases like Down's syndrome and hemophilia,[2] but such problems as cancer, heart disease, and aging[3] as well. Much of what is exciting in biology today—cloning, genetic therapy, DNA vaccines, mapping the human genome—harks back directly to this work.

For their part in the discovery three men—James Watson, Francis Crick, and Maurice Wilkins—were awarded the Nobel Prize in physiology or medicine in 1962.

The Nobel ceremony was no less stately and impressive than it ever is. In his presentation speech, Professor A. Engström of the Royal Caroline Institute discussed the profound effect that the discovery has had on all areas of biology. He highlighted the artistic nature of the molecule and the creative thinking of the three discoverers who were being honored.[4]

Hovering about, however, was the ghost of a young woman—a dedicated, even heroic, scientist who had died four years earlier, and who, some claimed, deserved at least as much credit as the three honorees.

It was, in fact, on the basis of Rosalind Franklin's highly specialized X-ray studies that the final discovery was made.

The few years leading up to the discovery could have seen a world-class scientific race between two teams: Watson-Crick at the Biophysics Unit of Cavendish Laboratory in Cambridge, England, and, at King's College, London, Franklin-Wilkins, both of whom worked in the same laboratory during this fateful period.

Could have, that is, if Franklin and Wilkins had been able to get along; the two of them might well have gone down in history as the discoverers of the double helix. What Franklin needed, several colleagues later observed, was a collaborator, which is exactly what her feud with Wilkins made impossible.

A Powerful Presence

Of all our chapters, save perhaps the one on Freud, the personal equation here is the most significant. Anne Sayre, a writer of fiction who knew Franklin well, has written the only full-length biography of her. She describes Franklin as having been a small, frail child, then adds, "possibly the strongest consequence of brief and early ill-health was that it set firmly her pattern of reacting to frustration with passionate indignation, of fighting every inch of the way rather than submitting for a moment."[5]

Some of this fighting spirit may also have come from the fact that as one of five children she was surrounded by brothers till age eight, when a second girl came into the family. A result here, says Sayre, was that she never developed "any exaggerated awe for the superior capacities of males." On the other hand, "as a grown woman she [Franklin] would sometimes refer to her youth as a period made tense . . . by her need to struggle for minimal recognition."[6]

The fighting spirit, Sayre adds, "never left her. Nor did the suspicion then acquired that it was disadvantageous to be a girl. How much, if any, immediate reality gave birth to this suspicion? This is extremely difficult, probably impossible, to determine . . . because what is dealt with here are such delicate shadings of psychological outlook . . ."[7]

Franklin was also brought up in a family tradition of strong argument but not hostility. Sayre says she often later "behaved in accordance with it, sometimes to the bewilderment of those who did not understand how it worked, and assumed that the only outcome of a sharp, hotly argued disagreement was either hostility or capitulation."[8]

Franklin's mother observed: "This combination of strong feeling, sensibility, and emotional reserve, often complicated by intense concentration on the matter of the moment . . . could provoke either stony silence or a storm. . . . But the strong will and a certain imperiousness and tempestuousness of temper, remained characteristic all her life."[9]

At St. Paul's Girls' School she received an excellent education. By age 15 she knew that she wanted to be a scientist, and majored in chemistry at Cambridge. According to Sayre, Franklin "possessed a fierce but happy sense of vocation."[10]

Her first real job took her to Paris, where she was assistant research officer at the British Coal Utilization Research Association (CURA). During her time there she was productive and happy. Her strong temperament was apparently not a problem, and her Gallic coworkers seemed to have no trouble with her somewhat argumentative style of discussion.

While there she worked on amorphous (noncrystalline) and inorganic materials, basically coal and graphite. After five years she began looking to apply her interests and capabilities to other materials, more specifically to biological materials that could be found or put into crystalline form, so that she might apply some of what she learned in this more complex area. And so she ended up at King's, where a very different atmosphere prevailed.

When she arrived at King's at the beginning of 1951, Maurice Wilkins was already well ensconced. She and he had certain things in common: they were close in age (he was four years older); both were British and had been educated at Cambridge University, though they had not met there, and both had solid middle-class backgrounds.

Wilkins was trained as a physicist. Later, he spent several years during World War II working on the Manhattan Project in the United States, then put in some time at the University of California on the separation of uranium isotopes by mass spectroscopy. Along the way he met, and then was divorced from, a young American wife. Like Franklin, he was moving over into biophysics.

But they were very different people in many ways. Horace Freeland Judson, whose massive 1979 summary of the DNA story[11] has become a standard in the field, writes of Wilkins: "Some who worked with him remarked that he could respond to vigorous disagreement only by turning aside."[12] We can easily see that Franklin might take this as being ignored, or even an expression of contempt on Wilkins's part.

Had they been able to get along, they—the physical chemist and the physicist, both looking to apply their respective backgrounds to the more enticing world of biophysics—could have learned much from each other.

DNA

Not long before Franklin was hired, Linus Pauling, an American chemist of wide repute, had described the physical structure of proteins. This understanding had proved extraordinarily helpful in many areas of science, and cleared up many of the chemical and biological puzzles surrounding these ubiquitous biological molecules.

It also started people wondering: proteins seem to do so many things in the body; could they also be the carrier of heredity? Some scientists thought so. Others suspected the real carrier was a different chemical substance also found in cells: deoxyribonucleic acid, or DNA.

Structurally, Pauling had found that the strength of the protein molecules lay in a helical spine (made up of alternating phosphates and sugars) that ran through the center of the protein, and that the spine supported a series of smaller molecules, called bases, that sort of stuck out like leaves or twigs from a branch. It was these that made proteins so interactive and biologically productive in so many ways. But this arrangement could not explain how our heritable characteristics are passed on so efficiently.

Among those who suspected the carrier was DNA, it was natural to begin with the idea that DNA also had some kind of helical structure. Knowing what we do now, it all seems so obvious. But if you can't see the structure, it's not obvious at all. The problem was that the DNA molecule, though large as biological molecules go, was still too small to be seen directly with any kind of laboratory instrument: too small for the optical microscope, and not able to survive the preparations needed for viewing by the electron microscope.

Somehow the structure would have to be deduced by other, more indirect, methods. The most promising was a technique that involved the use of X rays. A century earlier, Pasteur's work with crystals had already shown a connection between the crystalline form and the chemistry of a biological molecule. Early in the 20th century, Sir William Henry Bragg, knowing that X rays have a much shorter wavelength than visible light, found a way to use the periodic—that is, regular—placement of atoms in a crystalline material to bend, or diffract, X rays directed through the material. The emerging X rays would paint a distinctive pattern on a film plate, thus permitting the placement of the atoms to be deduced.

The overall study of crystal structure is called crystallography; the X-ray method, which has been considerably refined over the years, is referred to as X-ray diffraction, or X-ray crystallography. But it's a highly complex procedure, often requiring intricate calculations and, often, deep intuition.

Bragg's son, William Lawrence Bragg, joined his father in this work, and the two of them shared the 1915 Nobel Prize in physics for their discoveries. In 1938, Lawrence moved from the University of Manchester to head up the Cavendish Laboratory at Cambridge. The earlier director, J. D. Bernal, moved to Birkbeck College in London, taking with him, in the process, basically all the work that had been done to date with biological materials. Only Max Perutz, a young Austrian interested in proteins, remained at Cambridge. It was with Perutz that both Crick and Watson went to work.

And this is where Rosalind Franklin comes in, and why she was invited to join the King's laboratory. For both it and the Cavendish were part of an overlying group, the Medical Research Council. And the council, which wielded power on the basis of its grants, felt that King's had priority in the research on DNA, particularly in light of the fact that the Cavendish had built its considerable reputation on the basis of physics, not biology.

In 1946, when the King's lab began to emerge from its wartime activities, John Randall, a physicist, was brought in from the University of Birmingham to head up a new biophysics unit there; the objective was to apply physical methods to biological problems. Wilkins, who had taken his Ph.D. under Randall, was brought in under him.

Wilkins had done some work on crystallography as an undergraduate, but both Wilkins and Randall were basically physicists. Wilkins began with some work on the tobacco mosaic virus, a favorite of biophysicists, which involved just a bit of X-ray work.

So, aside from a few primitive attempts to apply X-ray methods to DNA, at King's and elsewhere, there was little in the field to show for it. Randall had gotten wind of Franklin's talent at using X-ray crystallography with inorganic materials. Looking for someone to apply X-ray methods to analysis of organic materials, and especially DNA, he invited her to come to King's and set up a laboratory with this as an objective.

Organizational Mayhem

Unfortunately, Randall never spelled out the situation carefully to either Franklin or Wilkins. Perhaps he hadn't thought it through carefully. Two historians suggest an even more troubling possibility. They suggest that Randall, whom they describe as "an ambitious, ruthless, no-holds-barred researcher himself," recognized the potential importance of DNA. "Could he," they ask, "have planned to isolate Franklin from Wilkins, so that she would report her findings directly to him? If so, he . . . might be the senior co-author (with her) of one of the century's most significant publications." After all, they add, "Randall had the reputation of running a tight ship, but he allowed the Wilkins-Franklin feud to fester for two years. . . ."[13]

In any case, she arrived thinking that the new X-ray crystallography unit she was putting together was her baby. For when she was hired by Randall, he told her in writing that "as far as the experimental X-ray effort is concerned there will be at the moment only yourself and [Raymond] Gosling, together with the temporary assistance of a graduate from Syracuse, Mrs. Heller."[14] It was not unreasonable for her to believe that the project was hers, and that Gosling and Heller would assist her. The term "at the moment" in his letter seemed unimportant to Franklin, but turned out to be significant.

When she arrived at the beginning of 1951, all pumped up and ready to go, she dived into the work. Wilkins, however, was busy with other things, and was often away, visiting other labs; but he always felt that when he was ready, there would be no problem with moving back into the X-ray work. In the fall of that year he was away again for several weeks, during which Franklin and Gosling generated some X-ray diffraction data.

When he came back, carrying some DNA samples he had gotten from a lab in New York, he assumed that they would then work together with the new samples. One of the pictures taken by Franklin and Gosling also looked interesting to him, but needed interpretation. He leaped into the work with enthusiasm and much talk about how it showed that DNA was helical.

As long as Wilkins had been busy elsewhere, there was no conflict. The problem was that Wilkins had apparently assumed all along that they would not only be collaborating, but that in some way Franklin would be reporting to him. Then again, was it only, as Crick suggests, "her suspicions that he really wanted her to be an assistant rather than an independent worker."[15]

In any case, Robert Olby, who has studied this history carefully, feels that she snapped at him, with something like, "Don't interpret my data for me!" [16] So nothing came of that data in the King's lab, but it was to provide critical evidence to the Cambridge group a year and a half later.

Remember that Franklin had recently come from a laboratory in Paris where she was not only treated as an equal, but where the atmosphere was one of free give and take, whatever the gender of those involved. If she barked when pushed or prodded, her antagonist barked back. There was no biting, however.

At King's and, apparently, at British labs in general at that time, this was not the way things typically worked. And, as noted earlier, Wilkins's personality compounded the problem. When conflict threatened, he simply backed away.

It was a bad start.

The Other Players

Francis Crick was at the time a research student working on his Ph.D. at Cambridge; he was doing some X-ray diffraction work with polypeptides (long molecule chains) and proteins under Perutz. But, basically a theorist, he hadn't the natural experimental feel that Franklin had.

James Watson, though 12 years his junior, already had his Ph.D. in genetics—had in fact gotten it at the impressive age of 22 at Indiana University. But although he had had some experience with phages (viruses that prey on bacteria), he also knew little or nothing about DNA. As for his presence in the Cavendish lab, he arrived in August 1951 and was basically a visitor seeking to learn something about genes. Interestingly, he was there on a fellowship originally funded by the National Foundation for Infantile Paralysis.

Crick and Watson met for the first time early in October. They were soon having lunchtime discussions at the Eagle, a picturesque but worn-down pub near the Cavendish. Both also suspected that DNA held some wonderful secrets.

Early Attempts

On November 21, Franklin and Wilkins presented a summary of their findings to date at a King's College colloquium. A fair amount of the data Watson and Crick later used were presented, mainly by Franklin, at that colloquium. Watson attended but, in his flamboyant, careless manner he took no written notes. And his level of understanding, which later increased rapidly, was at a fairly low level then.

Using Watson's reading of the King's group's presentation, Watson and Crick quickly came up with a model, and they invited the King's group, including Franklin and Wilkins, over to see it. Franklin, who by that time had generated some interesting data, blasted it, and apparently did not spare the model makers, either.

Watson and Crick nevertheless suggested that the two groups collaborate in the DNA work. The King's group spurned the offer. What could these two clowns—the gawky, skinny, ever-grinning 23-year-old Watson and the raucous Crick, whose laugh drove Sir Lawrence Bragg up a wall everytime he heard it—offer?

Bragg, in fact, was so disgusted by what happened that he told both Watson and Crick to stop all work on DNA. Ostensibly Crick returned to his doctoral dissertation on the alpha-helical models of protein structure, Watson to the genetics of viruses. Nothing had yet lit a real fire under the Cavendish duo, but both continued to mull the DNA problem.

Puzzle Pieces

Prior to Watson and Crick's discovery, there already was a fair amount known about DNA. It was a long molecule, rather than round or squat. It could, in fact, be drawn out of a solution as a long fiber. There was some sort of spine, as with proteins, and a set of four different bases: adenine, thymine, guanine, and cytosine. Scientists also knew the chemical makeup of the bases (what atoms they were made up of), and even that they were flat molecules, rather than curled up as some molecules are.

What was not known was how all the pieces fit together, or how they were held together. Did the bases connect to each other? How? Like to like? Like to unlike? Did the bases form a straight chain as the spine did?

But clues were starting to float about. In preparation for the November meeting at King's, Franklin had also written a progress report, which Watson and Crick got to see early in 1952. In it, Franklin suggested that the bases faced inward and were held in place by the phosphate backbones. It was to prove an explosive finding, though not right away.

There was another clue. The four bases seemed to pair off: adenine to thymine; guanine to cytosine. This had actually been shown in 1949 and published in 1950 by Erwin Chargaff, an Austrian-American biochemist. It was work of which Watson and Crick were blissfully unaware.

They learned of it in a curious way. In May 1952, Chargaff visited Cambridge and was introduced to Watson and Crick. In their discussions, he mentioned his finding and, expecting kudos, was astounded to find that neither knew of it.

After the meeting, they knew of it. And it turned out to be the first important piece of the puzzle to fall into place for them. For some reason, although they credit one of Chargaff's publications in their *Nature* paper, they did not refer to this work.

Chargaff's bitterness at this fact shows up clearly in his description of the meeting with Watson and Crick.

> The first impression was indeed far from favorable and it was not improved by the many farcical elements that enlivened the ensuing conversation. . . . The impression: one [Crick], thirty-five years old, the looks of a faded racing tout, something out of Hogarth. . . , an incessant falsetto, with occasional nuggets glittering in the turbid stream of prattle. The other, quite undeveloped at twenty-three, a grin more sly than sheepish, saying little, nothing of consequence.

His next sentence is pregnant with meaning: "I recognized a variety act, with the two partners at that time showing excellent teamwork. . . . They wanted, unencumbered with any knowledge of the chemistry involved, to fit DNA into a helix. . . ."[17]

His main objection, though, is that they had known nothing about his base-pairing work. In fact, Watson and Crick had even put Chargaff somewhat on the defensive. Crick: "We were saying to him as protein boys, 'What has all this work on nucleic acid led to? It hasn't told us anything we want to know.' "[18]

But after Chargaff told them of this work, Crick admitted, "Well, the effect was electric. . . ."[19] Yet he also argues that the key discovery, the exact nature of the two base pairs, was Watson's, and that "he did this not by logic but by serendipity . . . he was looking for something significant and *immediately recognized the significance of the correct pairs when he hit upon them by chance.* . . ."[20] It was the first glimmering of how the double helix does its work.

However it happened, the first piece of the puzzle was in place. In the same month, Franklin got an excellent picture of a DNA sample, which she studied for a bit and then apparently just put away in a drawer. It showed a specific X-shaped pattern (of the so-called B form) that was characteristic of a helical shape. Another piece of the puzzle had been created, but was temporarily buried. By this time, Franklin was so unhappy that she was beginning to search for another lab to work in.

Next Puzzle Piece

Still, DNA began to look more interesting all around. By the fall of 1952, Linus Pauling, he of protein model fame, was even preparing a paper on the DNA problem. In a strange coincidence, Pauling's son Peter was not only working at Cambridge but was sharing Watson and Crick's lab space.

André Lwoff, another minor actor in the drama, describes what happens next. Peter, he writes, "receives detailed letters from Pasadena and informs his colleagues of the evolution of his father's work, seemingly without telling his father what Crick and Watson are up to. Freud would have been interested in the situation."[21]

They, and especially Watson, began to feel the temperature of the search increasing. Watson, in fact, had moved into a race mode: he now saw the whole thing as a three-way race between King's, Cambridge, and now Pauling at Caltech in the United States. He knew that Pauling could be a powerful competitor.

Watson and Crick informed their bosses that Pauling was clearly in the race: See, everyone is working on DNA. You no longer have to bend to this silly rule that says King's has priority.

Crick also asked Wilkins if he would object to their playing a bit with some DNA models. Wilkins, who thought the solution would more likely come from his theoretical approach, offered no objections. Good thing, says Watson: "When Maurice's slow answer emerged as no, he wouldn't mind, my pulse rate returned to normal. For even if the answer had been yes, our model building would have gone ahead."[22]

Most important, Bragg did relent. Here, too, competition may have reared its head. For Pauling had several times made discoveries in areas in which the British, including Bragg, had considered themselves preeminent. This included the alpha helix work. Bragg didn't want it to happen again.

The race was on. But it was a funny kind of race. Watson and Crick were off and running, but never bothered to mention it to Wilkins, who in turn might very well not have mentioned it to Franklin even if he had been told. But by this time Franklin had made arrangements with another lab, and was planning to move over early the next year.

As for Pauling, who knows what might have happened if his passport had not been taken away in 1952 by a government suspicious of any political dissent. (Pauling was suspected of being a leftist, this in the glory days of Joe McCarthy.) He was on his way to London when it happened, and he might very well have gotten to see Franklin's picture. Fate continued to smile on the Watson-Crick team.

And then, in December 1952, they got another lucky break. John Randall hosted a presentation at King's for the Biophysics Committee, which consisted of some of the top brass in the Medical Research Council. The objective was to apprise them of progress made to date by both laboratories—in other words, of how and on what their grant money was being spent. Randall had a report prepared, part of which was written by Franklin and Gosling.

Among the attendees was Max Perutz, the distinguished chemist from the Cavendish group, who, as a result, then had in his hands the data that showed several important aspects, including a helical shape and certain conformational relationships for a specific form of DNA (the B form).

What happened then is a little shadowy, but Crick apparently heard about the report from Wilkins, with whom he was in frequent communication. And he, or perhaps Watson, asked Perutz if they could see the report.

Perutz let them see it. He realized later that he should have asked Randall for permission to show it to Watson and Crick. But, he wrote later, "I was inexperienced and casual in administrative matters and since the report was not confidential, I saw no reason for withholding it."[23]

A Fateful Visit

By this time, Watson was visiting the King's College laboratory on a fairly frequent basis. One of these visits, in January 1953, was to prove fateful for all concerned. Watson had with him a draft of Pauling's paper on DNA, obtained from Peter. Watson had found a major error in it and wanted to discuss it with Wilkins. Among other things, Pauling had proposed a three-chain model,[24] with the sugar-phosphate chain going down the center.

Arriving at King's, Watson found that Wilkins was busy, so he went on to Franklin's lab. As he described the situation later, the door to her lab was open, so he just walked in. Franklin, who had been staring intently at one of her X-ray photos, was startled and,

> looking straight at my face, let her eyes tell me that uninvited guests should have the courtesy to knock. . . .
>
> Though I was curious how long she would take to spot the error, Rosy [a nickname coined at Cambridge that no one ever used to her face] was not about to play games with me. I immediately explained where Linus had gone astray. . . . She became increasingly annoyed with my recurring references to helical structures. . . .

There were a couple of reasons for what happened next. One is that she probably was not too happy with being lectured to by this gawky, inexperienced young American. Second, she was beginning to question the very idea of the helix as the basis for the DNA molecule.

Watson continues: "Cooly she pointed out that not a shred of evidence permitted Linus, or anyone else, to postulate a helical structure for DNA." But the discussion was heating up: "Rosy by then was hardly able to control her temper, and her voice rose as she told me that the stupidity of my remarks would be obvious if I would stop blubbering and look at her X-ray evidence."

Watson in turn

> implied that she was incompetent in interpreting X-ray pictures. . . .
>
> Suddenly Rosy came from behind the lab bench that separated us and began moving toward me. Fearing that in her hot anger she might strike me, I grabbed up the Pauling manuscript and hastily retreated to the open door. My escape was blocked by Maurice [Wilkins], who, searching for me, had just then stuck his head through. . . .
>
> Walking down the passage, I told Maurice how his unexpected appearance might have prevented Rosy from assaulting me. Slowly he assured me that this might very well have happened. Some months earlier she had made a similar lunge toward him. They had almost come to blows following an argument in his

room. When he wanted to escape, Rosy had blocked the door and moved out of the way only at the last moment.[25]

I give this long quote for two reasons. First, the likelihood that Watson's reaction was exaggerated. Franklin certainly had no history of violence. Keep in mind that Wilkins's behavior is coming to us via Watson's pen. Horace Freeland Judson points out: "The scene would have been still more comical if Watson had reminded us that she was short and slim, he over six feet, if scrawny."[26] Judson calls the whole thing a "ludicrous encounter."[27] Nevertheless, the image is such that it creates a picture of a powerfully aggressive, masculine woman, one that easily serves the purpose of anyone looking for a way to blame the feud on Franklin.

Second, and more important, is what happened next. Watson continues:

> My encounter with Rosy opened up Maurice to a degree that I had not seen before . . . he could treat me almost as a fellow collaborator rather than as a distant acquaintance. . . . [Then he told me that] Rosy had evidence for a new three-dimensional form of DNA. . . . When I asked what the pattern was like, Maurice went into the adjacent room to pick up a print of the new form they called the "B" structure.
>
> The instant I saw the picture my mouth fell open and my pulse began to race. . . .[28]

This was the famous photo, number 51, that Franklin had taken in May, and which was to play a pivotal role in the eventual discovery. Ironically, Franklin may never have known that Watson saw it. She also didn't know that Watson and Crick had also seen earlier data that she had produced, which also suggested a two-chain helix, with the bases in the center!

Anne Sayre quotes Wilkins as admitting to her later on, "Perhaps I should have asked Rosalind's permission [to show the photo] and I didn't. . . ."[29]

Ironically, Wilkins himself, though physically located at King's, also had not seen the photo when Franklin first came up with it. So we have yet another "what if": what if he had seen it some eight months earlier when she took it? If the two had been working together as they should have been, they might very well have had the answer.

Confusion

Apparently it was not all gloom and doom at the King's lab. For about six months earlier, Franklin and Gosling had circulated a small black-bordered card on which she had written. "It is with great regret that we have to announce the death, on Friday 18th July, 1952, of D.N.A. HELIX (crystalline). Death followed a protracted illness . . . A memorial service will be held next Monday or Tuesday. It is hoped that Dr. M. H. F. Wilkins will speak in memory of the late HELIX."[30]

It was a cute way of announcing her turn away from a helical DNA, but it presaged many months of lost time—at least as far as finding the structure is concerned. The problem was that although Franklin was an accomplished crystallographer, she was uncomfortable with speculation. She wanted precise, specific data and was not willing to go beyond it. When she saw something unexplained in her photos, she couldn't ignore it. This characteristic had served her well in her earlier work with the coal group in Paris, but it made for trouble in her efforts in the more complex world of biology. It happened again when she came up with another excellent photo, but of a different, slightly drier, state of DNA, the so-called A form; its details were even clearer than those seen in her photo number 51 of the B form.

Prior to Franklin's wonderful facility with this technique, X-ray photos of DNA invariably showed a mixture of two phases, which had confused the issue for earlier workers. It was she who, using carefully controlled humidity conditions, discovered there were two phases, the B form being slightly more hydrated (damp).

The A form, being drier, actually gave more detail. But the greater detail turned out to be a monkey wrench in that it threw Franklin off—even to the point of feeling that the DNA molecule was not helical. Caught up in trying to analyze her photo of the A form, she devoted months to trying to tease out its details. In so doing, she actually set aside, at least temporarily, her own evidence that the structure of the B form was helical.

Ironically, the A photo's detailed complexity eventually turned out to be extremely useful—but that was later, when it helped corroborate the Watson-Crick discovery.

As for model building, she considered that that would be a waste of time until she had all the experimental data in hand. Wilkins, her potential collaborator, had done some model building in his earlier career, experience that could have been enough to allow Franklin-

Wilkins to crack the case. For as Crick points out: "DNA is, at bottom, a much less sophisticated molecule than a highly evolved protein and for this reason reveals its secret more easily. We were not to know this in advance—it was just good luck that we stumbled onto such a beautiful structure."[31]

Crick even says that "in our enthusiasm for the model-building approach we not only lectured Maurice on how to go about it but even lent him our jigs for making the necessary parts of the model."[32]

So the feud between Franklin and Wilkins had a double-barreled effect: it prevented her from interacting with one or more colleagues and perhaps discovering her error earlier; and it drove Wilkins into the willing arms of Watson and Crick.

Bingo!

Crick tells us, "Originally my view was that the X-ray diffraction patterns of the DNA fibers was a job for Maurice and Rosalind and their colleagues at King's College, London, but as time went on both Jim and I became impatient with their slow progress and their pedestrian methods. . . ."[33]

Happily for them, the pieces were rapidly falling into place. Watson's deeper knowledge about genetic function, honed by his earlier experience with phages, came in handy, as he was doing most of the model building; but he kept trying to put the "spine" on the inside. Again, the collaboration paid off. Crick finally persuaded Watson to try reversing the order: spine on the outside, bases on the inside. It was also Crick who discovered another major feature: the codings for the two sides of the structure are antiparallel (run in opposite directions)—"a feature I had deduced from Rosalind's own data." He also calls the specific pairing of the bases "a key feature of the structure."[34] The importance of their collaboration becomes ever more obvious.

Watson was so energized by all that was happening that he couldn't wait for the lab's model builders to finish their metal replicas of the various atoms in the DNA molecule. He cut out a batch of bases from cardboard and started trying to fit them in various combinations into the double spine. Suddenly the Chargaff ratios came into play and the whole thing made sense. They got a few things wrong, but basically the structure was up and holding. The final attack had only taken a few weeks.

Ironically, earlier on the same day in which the model was built, the mailman had brought a note from Wilkins saying that he was now clear of Franklin's presence and was about to begin DNA model building in earnest. Too late. Too bad.

The power of the structural approach in this case, as opposed to a theoretical, analytical method, is clear in Watson's next comment: "so we [Watson and Crick] had lunch, telling each other that a structure this pretty just had to exist."[35]

Crick remembers another lovely vignette: "I recall going home and telling Odile [his wife] that we seemed to have made a big discovery. Years later she told me that she hadn't believed a word of it. 'You were always coming home and saying things like that. . . . ' "[36]

They called Wilkins and Franklin to show them the model—the famous double helix. All saw quickly that the Cambridge team basically had it right, and that it meshed with Franklin's data. What they saw looked a little like a spiral-shaped zipper with rigid sides (the phosphate-sugar chains) and the bases as the teeth. These are joined, not mechanically as in a zipper, but by weak chemical (hydrogen) bonds. This permits the spines to "unzip" easily and make copies of itself, in this way passing on the information contained in its genetic code.

Wilkins, and also Franklin and Gosling, had been preparing papers on their work. It was jointly decided that their papers would appear in the same issue of *Nature* with Watson and Crick's, and that is the way it went.

Genuine Delight

When I first read, in several sources, that Franklin "showed genuine delight"[37] upon seeing the model, I couldn't believe my eyes. No one, I thought, was that saintly. Having worked so hard, and having come so close, how could she not be chagrined, even furious? Something was missing.

What had happened was that Franklin *was* delighted that their model meshed so well with her data—because she never realized that it was *based* on her data! Sayre says "she had no notion at all that anyone outside King's had access to her unpublished results, much less that anyone had used them. . . . Rosalind knew only that Watson had come to her seminar in the fall of 1951, that he had listened rather opaquely to her early results, and that he had not subsequently indicated the slightest interest in them. . . ."[38]

Watson admits that he spent at least part of his time at the presentation wondering what Franklin would look like if she removed her glasses and did something with her hair.[39]

More from Sayre: "She [Franklin] seems to have taken the Cambridge structure as it was presented, as a work of perception, insight, and inspiration, and though she was pleased that it confirmed her work precisely . . . she did not know that, indeed, it incorporated her work."

What if she had known? Sayre continues: "My own guess—freely disputable—is that Rosalind might well have risen like a goddess in her wrath, and that the thunderbolts might have been memorable."[40]

Wilkins was of more help to Watson and Crick than to Franklin in other ways, as well. Robert Olby points out, "It is true that Watson ruled out three-chain models for biological reasons, but not before Wilkins told him that the physical data on DNA did not exclude two-chain models."[41]

The Double Helix—*the Book*

Now the story takes another twist.

Some 15 years after the great discovery, James Watson's very personal account of the history, *The Double Helix,* appeared. It was perhaps the first record of a world-shaking scientific event that tried to show in lay, even chatty, terms what was going on behind the scenes. Watson's impetuosity, his powerful ambition, his ingenuous outlook, all shone through clearly. Not everyone liked the result; some were outraged; but the book fascinated all who read it.

Thanks to the book's huge popularity, it deluged the reading public with Watson's take on the situation. But there was something very special about this take.

Feeling in his heart that he and Crick were on to something big, he had recorded his impressions as they happened. He tells us that he had kept a rather detailed diary, and that he drew from it in creating the book. In trying to be honest and, perhaps, a bit outrageous, he wrote the book as if he were writing it when all of the action was actually taking place. In other words, rather than modifying his take on the years 1951–1953 in the light of a later change of heart, he recorded his impressions in the 1968 book as if they were taking place a decade and a half earlier. And it turns out that his feelings about Franklin at the time of the discovery were quite different from what he felt later on.

But the reader doesn't learn that till the last half dozen pages. There, he admits that as his initial impressions of her, both scientific and personal (as recorded in the early pages of this book), were often wrong, I want to say something here about her achievements. The X-ray work she did at King's is increasingly regarded as superb. The sorting out of the A and B forms, by itself, would have made her reputation; even better was her 1952 demonstration . . . that the phosphate groups [that is, the spines] must be on the outside of the DNA molecule.[42]

Later, when she moved to Bernal's lab, she quickly extended Watson and Crick's qualitative ideas about helical construction into a precise quantitative picture.

By then, he continues, "all traces of our early bickering were forgotten, and we both [he and Crick] came to appreciate greatly her personal honesty and generosity, realizing years too late the struggles that the intelligent woman faces to be accepted by a scientific world that often regards women as mere diversions from serious thinking."[43]

Unfortunately, this appears on page 226 of Watson's book. By that point the damage has been done. Though, with careful reading, the reader might have understood that Franklin was doing some extraordinary work, the information is buried under the picture of "Rosy," a woman who is self-absorbed, unfriendly, even physically aggressive, and who happened to take a couple of useful X-ray pictures.

The Feminist Critique

Anne Sayre, who died in 1998, was a novelist and short-story writer who happened to be married to a crystallographer. So, although not a scientist, she spent time in biomedical circles and eventually got to know Franklin well.

She was one of those who were outraged by Watson's book. Not only had Franklin been robbed (re the Nobel Prize), but now her image had been sullied as well. And so Sayre wrote her biography of Franklin: *Rosalind Franklin and DNA*. The subtitle, *A Vivid View of What It Is Like to Be a Gifted Woman in an Especially Male Profession*, shows that Sayre had more in mind than a straightforward biography. She refers, for example, to "a rationalization which implies that Rosalind, as an impediment standing squarely in the path of scientific progress, deserved to be pushed aside."[44]

Thanks largely to the book, though for other reasons as well,

Franklin became an important symbol of the feminist cause. In the *New Statesman,* for example, Jon Bate and Hilary Gaskin wrote: "Sayre's scrupulously researched and well-written book [tells of] the way in which women have been subtly displaced from the pantheon of science."[45]

But just as Watson's *The Double Helix* set the stage for a mistaken, or at least exaggerated, idea of what had taken place, so, too, did Anne Sayre's book. Perhaps, so outraged by the apparent injustices to Franklin, she never got to read the concluding section of Watson's book. Or maybe she felt, with some justification, that its impact was too little, too late.

She closes with a heartfelt statement that *The Double Helix* did damage to more than Franklin's memory. She fears that Franklin had been turned into some sort of ogre and that Watson's picture of her was being "used to menace bright and intellectually ambitious girls." She explains, "I once went to a public meeting of a local school board and heard a man stand up to demand that science requirements for girls be dropped from the high school curriculum because he had a daughter, and he 'didn't want her to grow up like that woman Rosy-what's-her-name in that book.' "[46]

Unfortunately Sayre, like Watson, went too far. I draw for this on two major sources. First, Crick, who actually became very friendly with Franklin in the remaining years of her life, admits that "there were irritating restrictions—she was not allowed to have coffee in one of the faculty rooms reserved for men only—but," he argues,

> these [restrictions] were mainly trivial, or so it seemed to me at the time. As far as I could see, her colleagues treated men and women scientists alike. . . . The only opposition I ever heard about was that of Rosalind's family. She came from a solid banking family who felt that a nice Jewish girl should get married and have babies . . . but even they did not provide really active opposition to her choice of a career.[47]

He adds,

> Feminists have sometimes tried to make out that Rosalind was an early martyr to their cause, but I do not believe the facts support this interpretation. Aaron Klug, who knew Rosalind well, once remarked to me, with reference to a book by a feminist, that "Rosalind would have hated it." I don't think Rosalind saw herself as a

crusader or a pioneer. I think she just wanted to be treated as a serious scientist.[48]

Judson is even more blunt about Sayre's reference to a male-dominated profession: "That view is false." He, too, admits that Franklin had a "formidably unpleasant time at King's," but, he says, a male-dominated profession was not the reason.[49]

His evidence is strong, and goes back to the 12-page report that Randall had distributed during the December 1952 meeting at King's. Of 31 listed scientists, 8 were women. Plus, there were 2 more women not listed. In other words, he writes, women "worked as scientists at every level from top to bottom and held between a third and a quarter of the professional posts."[50]

The reason for the confusion, Judson feels, is that "most hid their gender by initials."[51] Why would they have done this? The apparent answer is that there *was* some sort of gender discrimination in the world of science. But this seems to have been far less the case with crystallography than with other sciences, and it certainly was not the case in the laboratories headed by either of the Braggs, by J. D. Bernal (to whose laboratory Franklin moved when she left King's), or at Franklin's lab at King's.[52]

Further, Judson was able to track down seven of the women who had worked for Randall, all of whom "agreed that women at their laboratory were treated equally."[53]

One of them, Dr. H. B. Fell (later Dame Fell), was the unit's senior biological advisor. Judson quotes her as pointing out: "They were *both*, I think, rather difficult people."[54]

Nor should it be thought that everyone was hiding information from Franklin. When she showed the A form to Crick and said it proved that DNA couldn't be helical, he told her that she was wrong. In a working collaboration this could be important information. In her case she was merely annoyed. Crick admits that his manner may have been patronizing. But had she paid more attention to his input, and less to his attitude, she would have been better off.

There is one other angle wherein her being a woman may have influenced her behavior. When she turned, even if temporarily, away from belief in a helical form for DNA, part of the reason, Crick suggests, may have been that she "felt that a woman must show herself to be fully professional,"[55] meaning that she had to have airtight proof of anything she proposed, any idea she sent out, any deductions she made from her pictures.

Aftermath

Watson, Crick, and Wilkins all went on to successful, productive careers. Crick taught at Cambridge University for two decades and then moved over to do brain research at the Salk Institute in California.

Franklin, too, began to do some good work in her new crystallography laboratory at Birkbeck College. In addition to her work on plant viruses, such as the tobacco mosaic virus, she also began to study the structure of the polio virus. But the respite didn't last long. By the autumn of 1956 her health was already failing. Franklin died of cancer at the tragically young age of 37, working as long as she could. Her last three papers were published posthumously.

Her death robbed her of more than the rest of her life. Much of the credit that should have been hers, that might have been hers if she had been included in the Nobel list, was denied even to her memory, and even by such as Linus Pauling. In an article he wrote for the DNA anniversary issue of *Nature,* he credits Wilkins for the B form of the photographs.[56] And a British Museum DNA display initially mentioned Watson, Crick, and Wilkins, but not Franklin, until a colleague, Dr. Mair Livingstone, complained.[57]

Franklin died in 1958; the Nobel Prize was awarded in 1962. If she had lived, the Nobel committee would have faced a daunting task. In all cases, no more than three honorees have shared a prize. Would they have had to break precedent?

Watson's answer to the question of who would have gotten the prize if she had been alive: "Crick, myself, and Rosalind Franklin."[58]

And as long as we're playing "what if?" how about this one: several writers feel that if Franklin and Wilkins had been able to get along, if they had been able to pass the ball back and forth, there is a good chance they would have gotten the answer, and perhaps even before 1953.[59]

In that case, it's even more likely that Franklin could have gotten into the Nobel ring. My understanding is that the Nobel nominating committees tend to stand clear of anything that smacks of conflict. Perhaps if there had been no conflict, and if Franklin and Wilkins had done the trick, the prize could have been awarded in time for Franklin to have received it.

How important was that prize? Crick says, "Rather than believe that Watson and Crick made the DNA structure, I would rather stress that the structure made Watson and Crick. After all, I was almost totally unknown at the time, and Watson was regarded, in most circles, as too bright to be really sound."[60]

As we've seen in other chapters, feuds are not necessarily bad things, in the sense that the contestants may even be stimulated by the activity. Clearly that was not the case here.

CHAPTER 10

Gallo versus Montagnier

The AIDS War

At the end of the 1970s medical people believed that epidemic diseases were a thing of the past, at least in the developed countries. Another accepted "truth" was that a kind of virus found in animals, called a retrovirus, did not exist in humans. Medical science also "knew" that cancer was not caused by a virus in humans.

Over the next decade, each of these truths fell by the wayside. First, at the beginning of the 1980s, the medical world suddenly became aware that it was facing a new epidemic of unknown origin. Researchers saw quickly that it was transmissible from person to person, and also that its major effect was the breakdown of the affected person's immune system. This led to a motley and confusing array of symptoms and, as a result, it came to be known as acquired immunodeficiency syndrome, or AIDS.

From mankind's earliest days, no other disease has so quickly established itself as a volcanic, and worldwide, menace. In less than 20 years, 16 million people have died from it; in addition, 33 million more have been infected—virtually all of whom are doomed as well.[1] And as bad as the epidemic has been so far, the threat is even greater.

One reason is the disease's unusual quality of not bringing down the patient quickly. As opposed to the typical infectious disease like plague, typhus, or polio, with AIDS the victim may remain contagious (and not know it) for a decade or more; the result is an enormous cache of infection always on the ready to spread to others.

Intense and dedicated work by the biomedical establishment showed that the cause of AIDS was a virus—which has come to be called human immunodeficiency virus, or HIV. The discovery was made by two laboratories at about the same time and was tentatively announced in 1983. It should have been the occasion for pride and honor on the part of the discoverers, yet it became wrapped in controversy and unhappiness for years.

One of the discoverers, Robert Gallo, then at the National Cancer Institute (NCI; part of the National Institutes of Health, or NIH) describes the problem as "an acrimonious controversy involving legal, moral, ethical, and societal questions that soon spilled over into the world of scientific research and threatened to poison relationships between scientists, as well as between the research community and the general public."[2]

His major scientific opponent, a Frenchman by the name of Luc Montagnier, at the time an important member of the Pasteur Institute in Paris, is now the director of the recently formed Center for Molecular and Cellular Biology at Queens College in New York. He, too, is still involved in AIDS work.

In addition to charges from the competing French group, a variety of U.S. governmental agencies got involved as well, and their charges against the Gallo group included perjury, obstruction of justice, mail fraud, and conspiracy to defraud the government. All the charges were subsequently dropped, but the aftereffects have lingered on.

When I asked Gallo—who is still one of the country's leading AIDS researchers, and who now heads up the handsome new Institute of Human Biology at the University of Maryland in Baltimore—about his own reaction to the goings-on, he would say only that six good years had been torn out of his productive life. But the pained expression on his face said even more.[3]

Puzzling Beginnings

On June 5, 1981, a major U.S. research organization, the Centers for Disease Control (CDC), reported in its journal that in the previous seven months five young men, all active homosexuals, had come down with, among other things, an unusual kind of pneumonia.[4] What was strange was that this type of pneumonia had almost always been found in patients whose immune systems had previously been weakened by drugs or disease, which did not seem to be the case here. And so the initial feeling was that the problem with these five cases had more to do with their lifestyle. Each of the five claimed not to have had anything to do with the others; yet some sort of sexual transmission, not yet traced, seemed likely.

A month later the same CDC publication, *Morbidity and Mortality*

Weekly Report, carried a second report on 26 homosexual men who were afflicted with Kaposi's sarcoma, a most unusual skin tumor. This affliction had normally been seen in the United States mainly in Jewish men of Slavic descent, but also in Africa in children and adults of both sexes.[5] Kaposi's sarcoma had also been seen in patients who were on immunosuppressive drugs for such procedures as kidney transplants. But the new outbreak was noteworthy for its numbers, and also because the affected individuals showed evidence of other infections similar to the ones seen in the first group of homosexual men.

Among the most challenging aspects of AIDS is that the target is invariably the very cells that make up the victim's immune system. As the immune system weakens, the patient becomes prey to all kinds of opportunistic infectious agents that had been present in the body, but had been held under control until then.

By May 1985, 10,000 cases of AIDS had been reported, most of them resulting in death within two years of the diagnosis.[6] But the problem also surfaced in other parts of the world, and in a way that only deepened the puzzle. Doctors in France and Belgium noted that their African patients with AIDS had generally not engaged in the two suspected lifestyles of the American group—homosexuality and drug use (passing the disease via shared needles). The disease there was also found among females as well as males and seemed to be spread more by heterosexual than homosexual practices, especially among African prostitutes.

As the cases mounted, researchers in several labs began looking for a microbial causative agent, perhaps a virus.

Finding the Cause

Viruses in general are difficult to deal with, for several reasons. One is their tiny size: they easily pass through special filters used in microbiology to filter out bacteria and other offending organisms.

The AIDS virus adds a special kind of challenge. For a long time researchers believed that viruses used the same general method of making needed proteins that all other living things use. Typically, an organism's DNA produces an intermediary operator, a single-stranded messenger (ribonucleic acid, or RNA), which then coordinates the process of producing the myriad proteins needed to keep life going.

Viruses use this method, too, but Gallo, who began his research career in 1966, was by the 1970s already involved in research on a second, and different, type of virus. Because it could reverse the normal genetic process, it came to be called a retrovirus. Notably, the retrovirus's genome is made up of RNA, rather than the more typical DNA.

Gallo focused on these entities. A major break came in the early 1970s with the discovery of a special enzyme called reverse transcriptase (RT), which is used by the retrovirus to create viral DNA. This means that the retrovirus's genetic information is in the same form (i.e., DNA) as the host's gene structure and so can more easily take over the cell's operating machinery. In other words, it has found ways to confound the highly complex and usually effective biological immune system. As Gallo puts it, the viruses "appear to be welcomed into the trusting arms of the cell."[7] This can cause not only the death of the cell but, even before that, the reproduction of more viruses.

At first there was little outside interest in this work; who cares if a microbe causes disease in mice?

Still, it led to the important idea that the same might be true of cancer in humans. Gallo, in fact, found preliminary evidence for RT in certain human leukemia victims. And this led to the idea that a virus might lie at the heart of the AIDS epidemic. The enzyme was also seen as a footprint for a retrovirus.

But before the connection between AIDS and a retrovirus could be shown conclusively, researchers needed the ability to culture the viruses, to keep a strain alive outside the body. John Enders had done this for polio (see chapter 8); Gallo and his coworkers found a way to do it for human retroviruses—first for a type that causes an unusual leukemia and later for HIV. But at the time, as he puts it, "the door of the house of molecular biology may have been opened, but most of us could not see in to decide which room to enter."[8]

In fact, working with retroviruses turned out to be a nightmare. One basic problem is that they have no mechanism to correct errors that crop up when their genetic material is being duplicated, as most other creatures do. The result is production of many nonfunctional viruses, but also many that then present a bewildering variety of guises to the machinery trying to get rid of them—and to the researchers trying to pin them down. "Today," says Montagnier, "nine subtypes of the major group HIV-2 plus the new subgroup HIV-0 have been identified, but the list is far from complete."[9] (Compare this with the fact that it took only three types to sow confusion in the polio vaccine search.)

First Glimmers

Gallo had made a serious misstep when, in 1977, he mistakenly announced to the scientific community that he had discovered a new human leukemia virus. But it turned out to be an animal retrovirus that had contaminated his cell line, and his reputation, which seemed on the way up, sank. He determined to be more careful in the future.

That future arrived a couple of years later when he did find the first known human retrovirus; working with others, he showed that it did cause human leukemia, a form of cancer. He named it human T-cell leukemia virus, or HTLV; it was the first real evidence of a retrovirus that caused disease in humans. (T cells are one of the two types of fighting cells produced by lymphocytes, and constitute an important tool in the immune system's arsenal.)

He was particularly careful this time, and when he finally published the work in 1980, it earned him a Lasker Prize, the highest accolade in biomedicine short of a Nobel Prize. He found a second such virus in 1982, which caused a different form of leukemia, and the viruses were then referred to as HTLV-1 and HTLV-2.[10]

By 1983 the National Institutes of Health were under pressure from an increasingly impatient public. Knowing of the importance of the AIDS question, and feeling impotent, the NIH sought a way to increase its involvement. In April 1983, just under two years after the CDC's initial reports had surfaced, Gallo was enlisted in the battle.

A variety of suggestive findings led Gallo to suspect that AIDS was somehow linked to a retrovirus, one he believed would be related to HTLV.

He chose to have his laboratory at NIH zero in on this possibility—a brave and, some thought, foolhardy choice at the time. And in fact it turned out that the connection he made between AIDS and HTLV was wrong; but it was an idea that bore fruit.

Within only weeks of the start of the search, lab members found such an HTLV virus in the cells of several AIDS patients. He began to draft a paper that would bring attention to this new class of viruses as a possible cause of AIDS.

The French Connection

At about the same time, the French group was also delving into the matter, but from a different point of view. A group of Paris physicians interested in the disease hypothesized, for example, that if a virus was causing destruction of T cells, then higher levels of the virus would be seen in earlier stages of the disease, before many of the T cells were infected and/or destroyed. Dr. Montagnier's team, working at the Pasteur Institute, began to explore this idea.

They also found evidence of reverse transcriptase in tissue samples from a gay male patient who was in an early stage of the disease. As we know, RT is a marker for the existence of retroviruses and perhaps a factor in AIDS. Using electron microscopy, they photographed viral particles in the sample.

Montagnier contacted Gallo in 1983 with the news and Gallo, who was preparing two papers for *Science*, encouraged him to submit his paper to *Science* as well, with the idea that the papers from both teams would appear at the same time. But the time until publication was short, and in his haste to produce the paper, Montagnier neglected to include the needed abstract. Gallo offered to write it, and Montagnier agreed.

Gallo described the French virus as similar to his own—which was, in the light of how little was known at the time, reasonably accurate. The problem was that Montagnier's paper followed Gallo's two papers in the journal, plus another on the same subject by another American researcher. The unfortunate result was that Montagnier's input, which seemed merely to confirm Gallo's findings, was pretty much ignored.

In addition, by the time the papers appeared—May 20, 1983—and perhaps even earlier, Montagnier was already questioning these conclusions. For example, upon exposing their virus to HTLV antibodies, which should have fought the virus, they saw no such result. He began to feel that their virus, which they found in about 20 percent of AIDS patients, was a new, or at least undiscovered, retrovirus, and one that had no connection with HTLV.

Montagnier, trying to distance himself from the HTLV connection, came up with the acronym RUB, a rearrangement of the initials of the patient from whom the French virus was attained. Gallo, with the help of a *Science* editor, managed to persuade Montagnier to say that his virus was part of the HTLV family. Randy Shilts, whose book on the history of AIDS, *And the Band Played On,* became an international best-seller, states that Montagnier considers the name change "as one of the greater mistakes of his scientific career, the first step in the 'long tunnel of darkness' that lay ahead." [11]

Montagnier later called his entry LAV, for lymphadenopathy-associated virus. (Lymphadenopathy refers to swollen lymph glands, a condition that often precedes the onset of AIDS. To help keep things straight, we will continue to refer to Montagnier's virus as LAV, although the term did not come into use until later.)

Two Views

Now the two narratives begin to diverge a bit.

Hoping to get things straightened out, Montagnier maintains in his book on the history of the discovery, he and Gallo met in Paris in June 1983. Over dinner at a fish restaurant on the Left Bank, he writes,

> the discussion quickly heated up. I presented one argument after another. Gallo would have none of them: he maintained that LAV was a variant of HTLV. . . . However, he invited me to take part in a meeting of a task force he had put together at the NIH and asked me to bring a specimen of LAV so that his collaborators could analyze it. . . .
>
> Before leaving France, I had had a proposal for collaboration between NIH and the Pasteur Institute. . . . We had then immediately deposited our main isolates of virus at the National Collection of Microorganism Cultures [in France]. . . . However, after the meeting I was overcome with exasperation that no attention had really been paid to our new virus. Now there was no longer any question of working together on these terms. . . . [12]

Gallo sees things a little differently:

> In all our previous discussions, we had agreed that the molecular analysis of his isolate [the sample containing Montagnier's virus, which he had obtained from French AIDS patients] would be done by us in a collaborative effort with their lab. . . . Any of their isolates sent to us would be given the same treatment as our own isolates . . . and the results would be published as a collaborative effort. . . .
>
> Now Montagnier pulled me aside to say the collaboration was

off. Under pressure that "it must all be done in the Hexagon" [France], he said he had to change the plans.[13]

Whichever version comes closer to the truth, Gallo maintains that this was an early cause of their conflict.

The next episode came shortly thereafter, at the fall 1983 virology conference held annually at Cold Spring Harbor Laboratory (located in New York and directed by James Watson). Both sides summarized their arguments, and the search was still at a stage where there were valid arguments on both sides. Gallo's side, remember, had found evidence of HTLV in some of the isolates. Further, new human viruses do not surface often. It would therefore take some doing on Montagnier's part to convince the Americans of his belief that a new virus was involved.

Personal Factors

But, says Gallo in his own book summarizing his side of the story, major differences in their personalities, and perhaps even in national approaches to scientific meetings, now began to intrude. According to him, some scientific meetings are basically collegial, at times even gentlemanly; others are strongly competitive. He tells of an early meeting at which he was almost destroyed by challenging arguments from a respected higher-level colleague, Sol Spiegelman.

> Over time, however, I came to know Spiegelman as a friend and to value that first experience with him. It had taught me an important lesson. If science is one part inspiration . . . , and a second part tenaciousness (the willingness to stick to your ideas while others reject and denounce them), the third part is the proof, the long, hard benchwork Thomas Edison must have been thinking of when he described genius as 98 percent perspiration.[14]

Speaking of his own laboratory, he adds, "I admit that with tough questioning I have at times, unfortunately, even hurt the feelings of some of my own staff, especially those not used to my ways. I do it not because I am on an ego trip but because I believe in the value of vigorous debate. It is almost instinctive. . . . This kind of questioning . . . would become the second source of conflict between Montagnier and me."[15] For, he adds, the Cold Spring Harbor meetings were of the more competitive kind. He tells about one young researcher who almost fainted from fright at one of those meetings.

What Gallo was seeking was some sort of proof of Montagnier's contention that AIDS might have multiple causes, perhaps including HTLV, but also LAV, and perhaps other causes as well. Such proof was, at the time, simply not available.

Furthermore, says Gallo (with just a bit of haughtiness), "I did not fully understand the lack of technology available to Montagnier, so I did not consider that he might have failed to do the many things he might have done before coming forward with new claims because he simply did not have access to the proper resources or experience." [16]

Here's Gallo's take on what happened next: "Instead of first acknowledging . . . the importance of his [Montagnier's] new information and commenting on the value of his contributions, I proceeded at once to ask for more details about some of Montagnier's assertions, challenging his with questions. . . . Montagnier was clearly rattled by the questions and seemed particularly disturbed that they had come from me." [17]

As a way of explaining why this happened, Gallo then goes on to describe their very different personalities. In what sounds almost like a replay of the Franklin-Wilkins situation (see chapter 9), he continues:

> We are not alike in our style, as people or as scientists. He is quiet, almost formal, holding his own counsel when competing ideas are being presented. If he speaks at all, it is usually simply to ask a question. I love the rough-and-tumble of intellectual debate and usually welcome attacks on my own ideas (though I admit not always, and depending on the source), even though I know that at the moment I will be uncomfortable with them. . . . [18]

Gallo continues, "I have come increasingly to regret that the tone or spirit of my questioning was too aggressive and therefore misunderstood. It widened the growing chasm between the two labs that were making the greatest contributions toward eventually controlling AIDS." [19]

Randy Shilts puts their personality differences even more strongly: "Montagnier and Gallo were as dissimilar as two human beings can be, and each made the other vaguely uncomfortable. While Gallo was chummy, aggressive, and charismatic, Montagnier held himself aloof and was frequently described as doughty and patrician. Still, Montagnier recognized Gallo as a leader in human retrovirology. . . ."

Fortunately, in spite of the unhappy feelings, the two laboratories did continue to work with each other. But as to what happened next, it again, in typical "Rashomon" fashion, depends on whom you talk to.[20]

According to Gallo, the proof he had wanted from Montagnier was finally achieved from a sample Montagnier had sent—but at Gallo's lab! His group was finally able to confirm that France's LAV isolate did indeed contain an AIDS virus, uncontaminated.

But this finding cost them dearly. For one thing, they were accused of deliberately obtaining the LAV sample because they had nothing of their own to work with. "Nothing," claims Gallo, "could be further from the truth."

Moreover, he says, "our lab's confirmation of the French findings helped them more than it helped us. We lent credibility to their work, not the other way around. . . . It [the isolate] was sent," he adds, "because the French had published results regarding LAV, [and so] they were obligated to send it to us upon our request."

He adds, "Not for a moment did I think that someday it might look bad for us that the French had sent us their virus. Hadn't we sent them cell lines, HTLV reagents, and IL-2 [interleukin-2, a growth factor found by Gallo's group], as well as providing technical information and ideas as they had begun their work?"[21]

But now many samples were flooding into Gallo's lab from all over the world; one of the complexities that continued to confuse matters is that "HTLVs normally produce steady but low levels of reverse transcriptase, after several weeks of culturing; but in the case of the AIDS virus the RT level spikes [rises rapidly, then falls] . . . by the time most of the samples arrived from our medical collaborators and we started looking for RT activity, we had often missed the retrovirus peak period."[22]

Nevertheless, by end of 1983, Gallo and his colleagues felt they could reliably identify the AIDS virus. And by early 1984, after a desperately intense series of experiments, tests, and culturing of cell lines, the researchers were convinced they had found the causative agent for AIDS, which they named HTLV-3.

Though the name was eventually changed to HIV, the finding was a powerful one. But, as Gallo laments, "in the euphoria we were experiencing, none of us could have foreseen that what should have been an unqualified victory would become one of the most challenged and scrutinized discoveries in modern scientific history."[23]

In his opinion he now had the AIDS virus in hand. Though he had not compared it with the LAV virus, he was ready to do this in a collaborative effort. As it happened, by July 1984, Gallo's laboratory independently found that major aspects of his HTLV-3 and LAV were identical. Montagnier writes:

> Gallo phoned me to tell me this amazing result, while implying that I might have contaminated our LAV with the living virus that Sarang [M. Sarngadharan, a coworker of Gallo] had brought to our laboratory in May 1984. Gallo wrote later that I had no reaction to this. If we had used videophones he would have seen me literally leap out of my chair, on the verge of apoplexy! I retorted that if there had been any contamination, it could have happened only at his end. . . .[24]

Montagnier, continuing: "If [the viruses were] identical, [Gallo] should not have changed its name [to HTLV-3]. This is my main criticism of Robert Gallo. He later acknowledged his mistake in private conversations, and believes he has paid dearly for it."

Nevertheless, Montagnier adds, "Even after fifteen years these events still leave a bitter taste in my mouth, even though I am aware that the extraordinary publicity Gallo rallied around his virus reflected onto ours. . . ."[25]

This wording implies (perhaps not deliberately) that Gallo had orchestrated the publicity. Gallo counters: "I did 0 [*sic*] for publicity."

Gallo also points out: "We wished to compare the two viruses side by side, and co-publish the results. We were not supposed to produce Montagnier's virus and compare it with ours alone. His rules."[26]

Press Conference

Again, factors other than the purely scientific compounded the oncoming agony. We saw in chapter 8 how a press conference—there, the one on the Salk vaccine—could go awry. It happened again in this case.

By April 1984, not long after Gallo's finding, reports were already circulating that AIDS was caused by a variant of Gallo's HTLV, the recently discovered HTLV-3, and the U.S. government decided to hold a press conference, even before the papers on the new work were to be published. This is normally not considered an appropriate procedure. But the pressure on the administration—from AIDS groups; from the public at large, which was wondering where its research money was going; and from some members of Congress—was so strong that the health department chose this route.

So, on April 23, 1984, Margaret Heckler, then the U.S. secretary of health and human services, stood at a microphone and told a large group of reporters that the probable cause of AIDS had been found, and that it was a variant of a known cancer virus, and was called HTLV-3.

Ironically, a *New York Times* reporter, Lawrence Altman, had, shortly before, run a story saying that researchers in France had also found the AIDS virus, and that they called it LAV. Oddly, and to the dismay of the French, the story had little effect. As we noted earlier, an important reason may be that finding a cause and proving it are two quite different matters.

As for the press leaks, Gallo as well as a few others at NIH and NCI were blamed for them. He hotly denies any responsibility. In fact, he argues, he would have been foolish to do this. For one thing, both the leaks and the press conference could have had the effect of preventing some forthcoming articles by Gallo and his group from appearing in *Science* magazine.

Nevertheless, this charge led to a shouting match between the Gallo group and one of Heckler's top aides. After things quieted down, the scientists briefed Heckler and the conference proceeded.

The aftereffects of the press conference, while not quite on the level of the Salk presentation, were powerful anyway. And that suited the administration's purposes, for the conference had been set up at least partly to answer strong complaints that the Reagan administration was doing little to deal with the worsening AIDS situation. Here, said the presenters, was solid evidence of our activity.[27]

Gallo tells me that while a carefully prepared text had been distributed to the press, spelling out the limitations of the finding, the reporters seemed to hear only Heckler's announcement. Even worse, says Gallo, she had laryngitis that day, and kept her remarks short, confounding the issue even more.

For example, in preparation for the conference, she had asked Gallo about the possibilities of a vaccine. He says he answered with a guarded, "Oh, possibly in a couple of years." In Heckler's strangled voice it came out that a vaccine would be ready for testing in two years, and that a blood test could be expected in six months.[28] The researchers cringed; the media, of course, ran with it, and the damage was done. Gallo knew that the French would be furious with him.

When the British publication *New Scientist* got wind of the news, they contacted a member of the Pasteur Institute, Jean-Claude Chermann, for comment. Like so many others, Chermann assumed that Gallo was to blame for the whole thing (press conference and leaks as well) and was enraged. Later, he recognized that the momentum was just too great, that little of this was Gallo's fault, and eventually he and Gallo not only made up but became close friends.

Heckler did manage to point out that Gallo had succeeded in producing large quantities of the virus, which was needed for development of a blood test for HIV—badly needed to prevent those getting transfusions from being infected, a growing problem everywhere.[29] Yet even this was to be an additional cause of trouble later on.

In the *New York Times* a few days later, the editor pointed out that neither a blood test nor a vaccine had actually yet been developed, and that "what you are hearing is not yet a public benefit but a private competition—for fame, prizes, new research funds. . . . The commotion indicates a fierce—and premature—fight for credit between scientists and bureaucratic sponsors of research. Certainly," the editorial concluded, "no one deserves the Nobel Peace Prize."[30] Although a successful AIDS vaccine is still a dream, Gallo did have a blood test available in the promised time, and many lives were saved by it.

NIH officials felt no compunctions about the way things were working out. They argued that the French would never have found their LAV without Gallo's preparatory work. "It was Gallo's own comments about a possible connection between HTLV and AIDS, they point out, that led the French to even look for a human retrovirus in the first place."[31]

Though the press conference did jump the gun, the scheduled articles did appear in *Science* magazine shortly thereafter, on May 4. In them Gallo reviewed the evidence in favor of an infectious agent as the cause of AIDS—still not widely believed at the time. He also, significantly, reported that the definition of his HTLV had been changed. Now, he said, HTLV no longer referred to human T-cell leukemia virus, but to human T-lymphotropic retroviruses, thus broadening the acronym's coverage, and that this was the basic cause of AIDS.

This was clearly a redefinition, but a useful one, for it now defined a family of viruses having some connection with the all-important T cells of the immune system. He had not yet provided solid proof, but things were moving along.

Nevertheless, not long after, early in January 1985, the roof started to fall in on Gallo. Further investigation, both at his lab and elsewhere, was already showing that one of Gallo's HTLV-3 samples was indeed virtually identical with Montagnier's LAV. Two possible readings of this—either one devastating and destructive—were (1) that the similarity was the effect of accidental contamination of Gallo's viral cultures, or (2) that Gallo's lab had actually based its work on the LAV culture.

For in truth the French had indeed come up with their isolate even before Gallo did, though he had better evidence to show for his work. He also now had 48 isolates to work with.

The nightmare began. It developed slowly, then picked up steam.

The Mighty Press

The American press began to play a significant role in what happened next. To begin with, says Gallo, "the rift was seen in international terms. Much of the French lay press would support Montagnier. The American press would start off supporting the other side's view of things, and suspect me simply because I was government-based." (I find this an odd, even suspect, comment, yet it certainly seems to have been borne out in subsequent years.)

He continues:

> American coverage was sometimes based on the assumption that the French had fairly and disinterestedly put forward their case, while the Americans had been driven by motives of greed, aggressiveness and competition. The French press is known for its boosterism, the American press for its skepticism, which it sometimes carries to an extreme. . . . I became the outsider in the French press and a villain in some quarters of the American press.[32]

Here's an example. For years, AIDS had remained a buried secret: it was "their"—gays'—problem, not "ours." Now, even as AIDS began to be recognized as a wider problem, another "secret" also began to surface. At a meeting in New York City on February 8, 1985, puzzled reporters were hearing a scientific presentation.

Montagnier and Chermann were there. Chermann had planned to spell out the promise of HPA-23, a mineral molecule that seemed to have some effect on HIV; Montagnier was to give some more details of the genetic properties of his LAV. Gallo had been scheduled to be there but had canceled out.

Montagnier said, essentially, that the genetic sequence of HTLV-3 and LAV differed by only one percent. Dr. Joseph Sonnabend, moderating, pointed out: "They are identical to a degree that would not be anticipated with two independent isolates from the same family."

What? Doctors and reporters both—suspicious, disturbed—dug for the meaning of the statement. Science writer Donald Drake of the *Philadelphia Inquirer* finally put it bluntly: "Are you suggesting that Gallo swiped his virus from the French?" Sonnabend, trying to keep the discussion in tow, pointed out other possibilities: for example, that possibly "Montagnier swiped Gallo's virus, or we are dealing with a very strange coincidence" (i.e., cross contamination).

All finally seemed to agree, however, that there was little likelihood that the similarity was coincidental. Shilts writes: "Montagnier knew enough about the chronology of Gallo's discovery to be suspicious, though he never publicly made the accusation himself. . . . [T]he NCI prototype of HTLV-3, announced in April 1984, could have been grown out of the same cells the French had cultured in January 1983. If it had, this had the makings of a scientific scandal of immense proportions."[33]

Over the next year or so the two sides held to their positions.

Then, as it became clear that the AIDS virus was transmitted by infected blood and other bodily fluids, both the French and the American labs saw this as the serious problem it was—and worked harder to devise a test to indicate its presence before symptoms appeared. This was an extremely important development for several thousand people who depended on transfusions that had already been infected by tainted blood.

Gallo's group had such a test by 1985. The French authorities, unhappily, chose not to use Gallo's test, leading to the deaths of more than 300 hemophiliacs and others who were transfused during surgery. Many more who became infected are still likely to die.

Michael B. A. Oldstone, a prestigious virologist, says there were two factors behind the French decision—both, in hindsight, foolish. First, unhappy with the fact that the American development seemed to be getting the main publicity, the French wanted to develop their own test. Second, they wanted to sell blood products previously collected and processed, which was less likely to take place under a foreign test.

This led to both investigations and criminal trials, and to the conviction and jailing of four health care workers. "It is still not clear," Oldstone reports, "how high up in the French government the scandal goes."[34]

He adds that France was not alone, that similar hesitation and balking took place in countries like Japan and Germany, with similar results—and similar suspicion of governmental connections.

What complicated the issue was that the French had applied for a patent for a blood test based on LAV in December 1983, which was not granted by the United States (or in other countries), while the

American patent, filed later, was granted. This led to a long legal battle between the Americans and French, which continued until March 1987, when an agreement was finally signed by the directors of the Pasteur Institute and the NIH. Gallo and Montagnier were to be considered codiscoverers of the AIDS virus, and the patents for blood tests were to be the joint properties of the two institutions, which would share the royalties. (The French blood test had by then been brought to a satisfactory level.)

Both sides, interestingly, pay special tribute to Jonas Salk who, as we know, had been through his own kind of hell. Respected by both sides, he helped out in several ways. As Montagnier puts it: "In the heated discussions that preceded this agreement, many of our colleagues acted as moderators, and in this regard I would like especially to thank Jonas Salk for his help."[35]

In addition, acting as a sort of honest interlocutor, Salk came up with his own official history of the discovery of the virus, which also helped in crafting a solution. The uproar had been loud enough that the agreement had to be signed by the presidents of both countries.

The result was "an equal apportionment of royalties between institutions, but," Montagnier still complains,

> not among the inventors. The three Americans—Robert Gallo, Mika Popovic[36] and Sarang—received $100,000 annually. After 1985, the French scientists received nothing for several years because their royalty share was first used to pay off the legal expenses. Later, in 1991, when the first net positive monies started to come in, the financial share given the French inventors was far from equal to that given our American colleagues.[37]

As often happens, the outside world grabs hold of the action and wants in. Gays feared, not unjustly, that a test required for HIV could harm them in a variety of ways: employment, insurance, socially. This, curiously, hurt Gallo as well. Sheila Jasanoff writes in the *American Scientist:* "In the conflict over causation, AIDS activists who were skeptical about the role of the HIV virus (and profoundly suspicious of its American codiscoverer, Robert Gallo) picked up and magnified the voice of Peter Duesberg, the most influential scientist to hold out against the dominant HIV theory. . . ."[38]

In fact, there still remains much that is not known and, though few mainstream scientists agree with him, Duesberg continues to challenge the widely accepted AIDS/HIV connection. He argues that the

HIV virus is a "confounding variable," and that the basic cause of AIDS is drug use.[39]

Along the way, a few legislators, particularly Michigan congressman John Dingell, saw themselves as the conscience of the nation and went after Gallo, although nothing had shown conclusively that he had done anything wrong. Certainly there was never any evidence that he had deliberately misappropriated the virus. But the suspicions continued to flow copiously—spurred on by the ever-helpful press.

"Some of this writing and reporting has been adversarial, on occasion outrageously so . . . ," complains Gallo. "I and other scientists have come to believe that the popular press is not always a disinterested medium for such communication. . . . As for those of us in AIDS research, few scientists working in this area have not at one time or another found themselves criticized, shouted at, shouted down, ridiculed, or harassed. . . ."[40]

Among the leaders here was John Crewdson, a newspaper reporter, who, starting around 1989, began attacking Gallo in his paper, the *Chicago Tribune*. Congressman Dingell began calling for a government investigation of all charges, and this led to, among other problems, a serious probe by the NIH's Office of Scientific Integrity. This went on for years.[41]

Then, beginning in November 1991, the Office of Inspector General at the U.S. Department of Health and Human Services began investigating the question of whether Gallo and Popovic, one of his employees, had made false statements in government documents about their HIV discovery and in their patent application for the blood test. Gallo's lawyer called one of the group's reports "nonsensical," and part of the government's "endless harassment" of his client."[42]

Popovic's reputation, in the meantime, plummeted to the point where he was out of work for significant intervals over a three-year period. He finally moved to Sweden and worked there until called back again by Gallo when his new institute was brought into being.

Basically, the investigation came to naught, with the convened panel finally concluding that Popovic, whose English was on the weak side, had made some small, honest mistakes in his explanations. This led to a suspension of the investigation of Gallo as well.[43]

In fact, things seemed finally to be quieting down by mid-1994—until Crewdson stirred things up again with another set of charges in his newspaper, this time claiming that Gallo and Popovic had improperly used a Montagnier sample in 1983 to create their blood test. Again, these charges petered out.

And what of Crewdson himself, and his reporting? Jon Cohen and *Science* magazine analyzed his methods in 1991 and decided that he did not make many major errors in fact. But, they immediately add, "in some instances, Crewdson has omitted facts—[especially when] these facts didn't fit conveniently with his overall argument of wrongdoing by the Gallo laboratory. . . ." [44]

Unfortunately, however, this sort of reporting tends to stick. In the *Random House Webster's Dictionary of Scientists,* published in 1997, one still reads in the Montagnier section that "the original work of Gallo's team was shown to have been incorrect and possibly even fraudulent." [45]

Detective Work

Both sides sought all along to figure out what had happened. Gallo's group looked into the freezers for evidence. One piece that surfaced was a sample of Montagnier's LAV-Bru (the Bru designates the patient from whom it was obtained), which had been sent to them in 1983. On further analysis, it revealed a contaminant, designated LAV-Lai. Gallo concluded that the original isolate he used in his discovery had also been contaminated. [46]

Gallo explains it this way:

One of our 48 isolates was cross contaminated with the virus sent to me by Montagnier. Much was made out of that as if we did it deliberately, i.e., took it. But cross contamination in virology is common. Ironically and rather amazingly, years later Montagnier had to acknowledge that he too had cross contaminated his original isolate, i.e., another HIV (which grows better in the lab) from another patient contaminated his original HIV isolate. It was that one that had contaminated one of our cultures. . . .

Then Gallo makes what I consider an important point: "Secondly, early in 1983 some of our isolates were a mixture of HIV and HTLV because the patients were infected by both. It took six months to sort this out." [47]

The later American detective work prompted the French to go back to their own cache, where they found evidence of the same problem that had confounded the American results. At least some of their own samples of LAV-Bru were similarly contaminated, showing the extraordinary difficulty of preventing contamination by these viral masters of disguise. The different strains respond differently to antibodies, and so had added to the confusion.

Based on these later analyses, however, French attorneys now claimed that Gallo had merely "rediscovered" the French virus, and they thereby sought a larger share in the royalties. Realizing that the disagreement was actually straining French-American relations, as well as discouraging international scientific cooperation, the American administration agreed in 1994 to an adjustment in royalties.

All along the way, as we have already seen, Gallo was subjected to an astonishing array of charges, all of which were eventually dropped. The result, unfortunately, was that instead of being able to concentrate on his own work, days and weeks at a time were instead required for combing records, for confronting accusers, for creating detailed paper paths.

Gallo says the experience would have been unbearable without the considerable support of colleagues at the National Institutes of Health and elsewhere.

Did Gallo learn anything from his agonizing experience? "I believe I've become more empathetic. I'm better able to identify with other people's feelings and problems."[48]

Montagnier, too, sees both good and bad in the competition between the labs. He writes: "The positive aspect of it was that researchers were stimulated to win the 'race.' The harmful aspect was the negative image it gave the scientists in the public's eye."

Today, he adds, "I consider the problem as over. We have a cordial relationship. . . . Strictly speaking, we do not have scientific collaboration, but I am regularly invited to his lab meetings in Baltimore."[49]

Perhaps, some day, the experience of both men—and the outrageous response of many outsiders—will pay off by preventing another such incident from hindering a major, ongoing research project.

But don't bet on it.

EPILOGUE

Many years ago my wife and I visited the magnificent old walled center of Dubrovnik, in what was then Yugoslavia. Looking out our hotel window before going to sleep we saw a strong red glow off in the distance.

The next morning I finally found someone who spoke some English and asked whether there had been a big fire in the city. She asked, "Is it still burning?" I said I didn't see any smoke, so I supposed not.

She shrugged. "So what do I care if there was a fire?"

If it didn't directly concern her, in other words, she had no interest. But health problems concern everyone, and so, as we saw in several chapters in this book, medical disputes may bring in others who might have done better to take on the pragmatic view of my Dubrovnik neighbor.

This is especially important for those who can throw their weight around—that is, to understand that there are inherent limitations on how accurate judgments can be at any stage of a scientific dispute.

Some controversies, for example, simply cannot be settled, or even understood, at too early a stage. The basic science may simply not be developed enough to permit sensible decisions; sometimes the very rules of the game may not be clear; there may not even be agreement on what evidence is relevant.

Nor can we assume that feuds can always be handled from a purely rational, scientific point of view. External factors—religion, nationality, social status—as well as personal ones—pride, greed, ambition—may play important, if not always obvious, roles.

The complexities inherent in these feuds show up, too, in the fact that resolution may not come even after the feuders are no longer around. Consider the revisionist science historians such as Nuland, Geison, Sulloway, and many of the writers who are still going after Freud. Even though I tend to disagree with the thrust of much of their work, we must carefully distinguish between what seems to me to be honest disagreement with the target's work, and writers (and officials) whose vitriol seems aimed at the target in a personal sense.

There's another point I'd like to emphasize. In the Introduction, I suggested that Leopold Auenbrugger might have been better off if he and/or his method had been attacked rather than ignored. What I've learned from my own research into the cases in this book, however, is that although medicine may profit from such attacks (in the sense of bringing things out into the open), those being attacked may very well pay a high price for their beliefs, discoveries, and new ideas.

Harvey was on the defensive from day one; Galvani died a bitter and broken man; Semmelweis ended up dying pitifully in a mental institution; Bernard suffered unbelievable mental agony, and lost his family as well; Cajal suffered a variety of what were almost certainly emotionally based illnesses; and Freud felt beleaguered during his whole professional life. Finally, we can't be sure, but although Pasteur seemed to thrive on controversy, his illnesses may very well have had a direct connection with the pressures and unpleasantnesses he faced; and Gallo told me point-blank that there were times during his siege period when he thought about jumping in the Potomac River.

Although Montagnier did not take the beating that Gallo did, he seems just as disgusted by what happened. He writes: "I learned that when scientific discoveries and feuds are highly covered by the Media, you have no friends. We say in French, 'Gardez vous de vos amis comme de vos ennemis,' which means, 'Watch your friends as well as your enemies.' This is very true, and I have seen around me jealousy, animosity, ingratitude and stupidity more than consideration and encouragement."[1]

For this alone, those of us in the outside world should, even when faced with what may seem at first to be devilment or stubbornness, at least try to be patient, understanding, and appreciative. After all, our lives depend on the researchers' success.

NOTES

Introduction

1. Auenbrugger, 1761, n.p.
2. Inglis, 1965, p. 127.
3. Debus, 1970, p. 101.

1. Harvey versus Primrose, Riolan, and the Anatomists

1. There have been two major translations into English *(On the Motion of the Heart and Blood in Animals)*. The latest is by K. J. Franklin in an Everyman's Library Edition (New York, 1963). I prefer an earlier one by Robert Willis (London: Sydenham Society, 1847), which has been republished by Prometheus Books (Buffalo, N.Y., 1993). All following citations refer to this edition. Portions of this translation can also be found in Eliot, 1910, pp. 59–139; Clendening, 1960, pp. 152–169; and Boynton, 1948, pp. 489–508.
2. Robinson, 1929, p. 159.
3. Richard Toellner, in Debus, 1972, p. 78.
4. *DMC*, p. 9.
5. Ibid., p. 24.
6. Ibid., p. 50.
7. White, 1965, p. 312.
8. Dick, 1957, p. 131.
9. Keynes, 1966, p. 24.
10. Ibid., p. 20.
11. Sir Thomas Barlow, "Harvey, the Man and the Physician," *British Medical Journal* 1 (1916), pp. 1264–1271, in Franklin, 1961, p. 47.
12. Bylebyl, in Gillispie, 1972, p. 151.
13. Dick, 1957, p. 131.
14. Debus, 1970, p. 101.
15. Majno, 1991, pp. 405–408.
16. Giovanni Argenterio, *De Somno et Vigilia Libri Duo* (Florence, 1556), p. 87, quoted in Bylebyl, 1979, p. 53.
17. Bylebyl, 1979, p. 41.
18. Majno, 1991, p. 395.

19. In his treatise *On Natural Faculties,* vol. 3, p. 15, quoted in Whitteridge, 1971, p. 45.

20. Thomas, 1980, pp. 132–133. Thomas was actually referring to the latter part of the 19th century in his reference to scientific medicine, but the point is certainly valid for earlier periods as well.

21. *DMC*, p. 23.

22. Ibid., p. 25.

23. Sigerist, 1971, p. 141.

24. In Bylebyl, 1979, p. 19.

25. *DMC*, pp. 51, 52.

26. Hamburger, 1992, pp. 39, 40. Hamburger (deceased 1992) described his book as an "Imaginary Journal." I don't believe this; I feel he had found Harvey's actual journal and for some reason refused to admit it.

27. This dedication was apparently a special one for the king and does not appear in the Philosophical Library edition. It can be found in Eliot, 1910, p. 61; part of it is given in Clendening, 1942, p. 152.

28. See, e.g., Gorham, 1994, pp. 211 ff.

29. Keynes, 1966, p. 452.

30. *Exercitationes et Animadversiones in Librum de Motu Cordis et Circulatione Sanguinis (Adversus Guilielmum Harveum),* 1630. His name, too, appeared in Latin, Jacobus Primirosus, with the result that in English translations the name is spelled either as Primrose or Primerose. I'll use the former.

31. Robert Willis, *William Harvey: A History of the Discovery of the Circulation of the Blood* (London: C. K. Paul, 1878), p. 213, quoted in Keynes, 1966, p. 320.

32. *Exercitationes,* 1630, quoted in Chauvois, 1957, pp. 222–223, and in Hamburger, 1992, p. 221.

33. French, 1994, pp. 192, 212.

34. Whitteridge, 1971, p. 255.

35. Also spelled Oluff Worm (French, 1994, p. 154).

36. Hamburger, 1992, p. 241.

37. French, 1994, pp. 233, 234.

38. In Keynes, 1966, pp. 232, 233.

39. Ibid., p. 233.

40. Ibid.

41. Ibid., p. 238.

42. Trent, 1944, p. 318.

43. Hamburger, 1992, p. 249.

44. Ibid., p. 175.

45. Mani, 1968, p. 130.

46. Ibid., p. 126.

47. Ibid., p. 130.

48. *DMC,* p. 16.

49. Paul Marquard Schlegel (also Slegel), an early supporter in Germany, tried, unsuccessfully, to convert Hofmann in 1638 (Keynes, 1966, p. 453).

50. Franklin, 1961, pp. 120, 121.

51. *Exercitatio Anatomica de Circulatione Sanguinis* (Cambridge, Eng., 1649). A translation by Kenneth J. Franklin, *On the Circulation of the Blood and Other Writings,* was published in 1953 (New York: Everyman's Library).

52. Frank, 1980, p. 33.

53. Franklin, 1961, pp. 103, 104.

54. Ibid., p. 108.

55. Hamburger, 1992, n. 38, p. 225. (Personal communication to Hamburger.)

56. Keynes, 1966, pp. 425, 426.

57. A. Cowley, *Verses Written upon Several Occasions* (London, 1663), pp. 18–21. Also in Keynes, 1966, pp. 427–429.

58. Keynes, 1966, p. 412.

59. Ibid., p. 421.

2. Galvani versus Volta

1. Otto von Guericke, a German physicist, built a rotating friction machine in the mid-17th century, the first device able to generate electrical charge on a continuous basis.

2. Abbé Nicholas Bertholon, who carried out physiological experiments with electricity. In Pera, 1992, p. 22.

3. Kellaway, 1946, p. 131.

4. Walsh, 1773, pp. 461–477.

5. Hoff, 1936, p. 163.

6. Aloisio or Aloysii is a Latinized Luigi, which is the form commonly used.

7. Egon Larsen, *Ideas and Invention* (London: Spring Books, 1960), pp. 246, 247.

8. Egon Larsen, *A History of Invention* (New York: Roy Publishers, 1961), p. 50.

9. Larsen, 1960, p. 247.

10. Static electricity is the same sort of electrical phenomenon that gives you a shock when you walk across a wool rug on a dry day and then touch an object that is electrically grounded. The shock takes place when the charge that builds up on your body is instantly discharged to ground.

11. Pera, 1992, p. 63.

12. Buckley, 1881, p. 259.

13. From their definitive publication, "A Bibliographical Study of the Galvani and Aldini Writings on Animal Electricity," published in the *Annals of Science,* 1936, vol. 1, p. 242.

14. Pera, 1992, p. 53.

15. Ibid., p. 86.

16. This and the immediately following quotations are from Margaret Glover Foley's translation (Norwalk, Conn.: Burndy Library, 1953), pp. 59–60. I took them from Mauro, 1969, pp. 141–142.

17. Volta, 1793, p. 286.

18. Ibid., pp. 288, 289.

19. De Santillana, 1965, p. 86.

20. Trumpler, 1994, pp. 701, 702.

21. In Pera, 1992, p. 183.

22. De Santillana, 1965, p. 87.

23. Arago, 1875, p. 129.

24. Giorgio de Santillana wrote: "Such was the setting for one of the great decisions of history, a decision that involved nothing less than the turning point of the scientific revolution" (1965, p. 86).

25. Kellaway, 1946, p. 130.

26. Pera, 1992, p. 170.

27. Dibner, 1971, p. 13.

28. Pera, 1992, pp. 119, 121.

29. Fulton and Cushing, 1936, p. 250.

30. Silver, 1998, p. 127.

31. Arago, 1875, p. 132.

32. Better known today as bioelectric potential, for until the circuit is closed, the electricity remains just that, a potential force waiting to be released.

33. Geddes, 1971, p. 45.

34. De Santillana, 1965, p. 89.

35. Heilbron, 1979, p. 77.

3. Semmelweis versus the Viennese Medical Establishment

1. Risse, 1975, p. 294, and Sigerist, 1971, p. 354.

2. Haggard, 1929, p. 66.

3. A Professor Dietl, quoted in Robinson, 1929, pp. 625–626.

4. Sigerist, 1971, p. 355.

5. Porter, 1998, p. 295.

6. In Haggard, 1929, p. 88. Holmes's paper on the subject, "The Contagiousness of Puerperal Fever," first appeared in 1843 in the *New England Quarterly Journal of Medicine*. It can be found in volume 38 of the Harvard Classics (*Scientific Papers;* New York: P. F. Collier and Son, 1910), pp. 223–254.

7. Robinson, 1929, p. 635.

8. I found this idea in Morton Thompson's novel, *The Cry and the Covenant* (1949), pp. 256, 257. Perhaps he made it up. But based on the enormous amount of research he did for the book, I'm betting that it reflected a true event.

9. Dr. Constance E. Putnam, at the Wellcome Trust in London, has made a study of Semmelweis and is working on a biography of him. She suspects that his forebears may well have been Jewish. Personal communication, March 30, 2000.

10. Robinson, 1929, p. 621.

11. Ibid., p. 613. For some more background on these men, see also Sigerist, 1971, pp. 291–302.

12. In Wangensteen, 1970, pp. 584, 585.

13. Robinson, 1929, pp. 634, 635. There's a bit of confusion here: Carter's translation of the *Etiology* (1983) has Semmelweis reporting that Michaelis had examined a cousin, not a niece, and that it was *after* she gave birth (pp. 177, 178). Both agree on the awful outcome, however.

14. Edwin Emerson Jr., *A History of the Nineteenth Century, Year by Year*, vol. 2 (New York: P. F. Collier and Son, 1900), p. 1110.

15. Weissman, 1997, p. 124.

16. Slaughter, 1950, p. 174.

17. Ibid., pp. 176, 177.

18. Carter, 1985, p. 35.

19. Semmelweis/Carter, 1983, pp. 246–249.

20. Ibid., p. 249.

21. Ibid., p. 208. Note that Carter, both here (p. 191) and in the biography (Carter and Carter, 1994, p. 64), places Scanzoni at the University of Prague. Nuland (1979, p. 268) and Sigerist (1971, p. 357) locate Scanzoni in Würzburg. Scanzoni apparently had a connection with both places (Semmelweis/ Carter, 1983, p. 212).

22. Semmelweis/Carter, 1983, pp. 232, 233.

23. Carter and Carter, 1994, pp. 88, 89.

24. Nuland, 1979, p. 268.

25. Ibid., p. 268.

26. Slaughter, 1950, p. 181.

27. Vaginal discharge that takes place in the week or two after childbirth.

28. Wangensteen, 1970, p. 579.

29. Carter, 1985, p. 50.

30. Ibid., p. 43.

31. Nuland, 1979, p. 267.

32. In Nuland, 1988, p. 256.

33. Haggard, 1929, p. 77.

34. Weissman, 1997, p. 124.

35. Carter and Carter, 1994, pp. 90, 91.

36. In Robinson, 1929, p. 624.

37. Carter, 1985, pp. 49, 50.

38. Robinson, 1929, p. 644.

39. See for example, "The Henry E. Sigerist Issue" of the *Journal of the History of Medicine and Allied Sciences,* April 1958, vol. 13, no. 2.

40. Sigerist, 1971, p. 358.

41. Semmelweis/Carter, 1983, p. ix.

42. Inglis, 1965, p. 154.

43. Nuland, 1979, p. 269.

44. Ibid., p. 270. Also in Nuland, 1988, pp. 260, 261.

45. Nuland, 1979, p. 267.

46. Nuland, 1988, p. 238.

47. *Book Review Digest 1949* (New York: H. W. Wilson), p. 913.

48. Semmelweis/Carter, 1983, p. 250.

49. Carter, et al., 1995, pp. 258–261.

50. Ibid., pp. 255–270.

51. Ibid., p. 268.

52. Carter, 1985, pp. 44–45.

53. Personal communication, March 30, 2000.

54. Carter, 1985, p. 36.

55. Ibid., pp. 46, 53.

56. For a discussion of modern terminology, incidence, treatment, and mortality (approximately 3 per 100,000 births in the United States), see, e.g., Carter and Carter, 1994, pp. 97–114.

4. Bernard versus Chemists, Physicians, and Antivivisectionists

1. Editor, 1878, p. 742.

2. Vogt, 1878, p. 23.

3. Sir Michael Foster, quoted in Wilson, 1914, p. 567.

4. Editor, 1878, p. 743.

5. Schiller, 1967, p. 249.

6. From Bernard's own notes for a course he was giving in his own laboratory. In Holmes, 1974, p. 344.

7. Schiller, 1967, p. 253.

8. *De la physiologie générale*, 1872, p. 203. Also in Foster, 1899, p. 164.

9. Foster, 1899, p. 166.

10. Robin, 1979, p. 23.

11. Bernard's words, from his *De la physiologie générale*, 1872. In Foster, 1899, p. 169.

12. Olmsted and Olmsted, 1952, p. 46.

13. Foster, 1899, p. 170.

14. Olmsted and Olmsted, 1952, p. 112.

15. Foster, 1899, pp. 169, 170.

16. Robin, 1979, p. 5.

17. Grmek, 1970, p. 29.

18. In Holmes, 1974, p. 430.

19. Schiller, 1967, pp. 257, 258.

20. Quoted in Olmsted, 1935, p. 353.

21. Ibid.

22. The field continues to develop. Bernard regarded the extracellular fluid as the internal environment. Later studies have shown that biological regulation also takes place at the cellular and even subcellular levels. See "Limits of the Internal Environment," by E. D. Robin, in Robin, 1979, pp. 257–267.

23. In Schiller, 1967, p. 252.

24. For more on the time factor, see "Limits of the Internal Environment," by E. D. Robin, in Robin, 1979, pp. 263–265.

25. Inglis, 1965, pp. 130, 131.

26. Bernard, 1957, p. 151.

27. Ibid., p. 82.

28. Hoff, 1982, p. 178.

29. Bernard, 1957, p. 15.

30. Ibid., p. 146.

31. Ibid., p. 9.

32. Tattersall, 1997, p. 374.

33. In Holmes, 1974, p. 447.

34. In Robin, 1979, p. 3.

35. Quoted in Holmes, 1974, p. 445.

36. Holmes (ibid., p. 135) describes Mialhe as "an able and experienced chemist." Joseph S. Fruton, in Robin, 1979 (p. 37), calls him "a pharmacist associated with the eminent chemist J. B. A. Dumas." No matter. The result was the same. One possible reason for the confusion is seen in Robinson's comment, "Like many of the builders of chemistry, Dumas began life as an apothecary . . ." (1929, p. 681).

37. In Holmes, 1974, p. 118.

38. For a recent discussion, see Bateson, 1992.

39. Porter, 1996, p. 164.

40. Editor, 1877, pp. 363–366.

41. Schiller, 1967, pp. 254, 255.

42. Darwin, 1958, p. 304.

43. Ibid., p. 306.

44. I list 19 references in my bibliography. For the most balanced accounts, see especially Aldhous, 1999; Bateson, 1992; Hampson, 1979; and special issues of *Scientific American* (February 1997, pp. 79–93) and *CQ Researcher* (August 2, 1996, pp. 673–693).

45. See, e.g., Kolata, 1998.

46. Hoge, 1999, p. 10.

47. See, e.g., Aldhous, 1998, p. 60; Hoge, 1999, p. 10; Bowden, 1982, p. 682.

48. In Olmsted and Olmsted, 1952, p. 241.

5. Pasteur versus Liebig, Pouchet, and Koch

1. *Judge*, December 12, 1885, vol. 9, no. 217, p. 2.

2. *Judge*, January 2, 1886, vol. 9, no. 220, p. 10.

3. Stephen Paget, in the *Spectator*. Quoted in Geison, 1995, pp. 265, 266.

4. De Kruif, 1938, p. 5.

5. *Encyclopedia Americana*, 1996, vol. 21, p. 516.

6. Geison, 1995, p. 49.

7. Ibid., p. 47.

8. Ibid., p. 39.

9. Dubos, 1976, p. 74.

10. Grant, 1959, p. 88.

11. Debré, 1998, p. 35.

12. Grant, 1959, p. 103.

13. Metchnikoff, 1939, pp. 22, 23.

14. Pasteur, 1879, in Eliot, 1892, p. 352.

15. Debré, 1998, p. 87.

16. Grant, 1959, p. 106.

17. Vallery-Radot, 1926, pp. 82, 83.

18. Debré, 1998, p. 103.

19. Sometimes spelled Marcellin.

20. De Kruif, 1926, p. 102.

21. Robinson, 1929, p. 688.

22. Pasteur, "On the History and Theories of Spontaneous Generation," from *Memoirs on the Organic Corpuscles Which Exist in the Atmosphere*

(1862), in Clendening, 1960, pp. 378–388. For a more up-to-date discussion, see Hellman, 1998, pp. 63–79.

23. Roll-Hansen, 1979, p. 279.

24. Farley and Geison, 1974, p. 189.

25. Grant, 1959, p. 124.

26. Robinson, 1929, p. 695.

27. De Kruif, 1926, p. 146.

28. Robinson, 1929, p. 701.

29. Ibid., p. 701.

30. Debré, 1998, p. 412.

31. Robinson, 1929, p. 701.

32. Alerting the body's immune system to a potential or actual threat, so that it has time to build up the body's defenses, e.g., specific antibodies to a disease-causing microbe, before being overwhelmed by a bacterial or viral onslaught.

33. Hendrick, 1991, p. 473.

34. Geison points out (1995, pp. 18–21) that one reason the notebooks have not been widely available has to do with the contretemps over Bernard's notes, which prompted Pasteur to instruct his family to protect the privacy of his own notebooks.

35. Ibid., pp. 4, 5, and dust jacket.

36. Ibid., p. 16.

37. Ibid., pp. 9, 10.

38. Ibid., p. 130.

39. Ibid., pp. 147–176.

40. Ibid., p. 12.

41. Ibid., p. 16.

42. Dixon, 1995, p. 46. See also chapter 7 in this book (on Freud), and the story of anthropologist Derek Freeman's attack on Margaret Mead, in Hellman, 1998, pp. 177–192.

43. Porter, 1995, pp. 3, 4.

44. Geison, 1995, p. 16.

45. Fee, 1995, p. 884.

46. Perutz, 1995, p. 54.

47. Geison, 1995, p. 17.

48. Perutz, 1995, p. 54.

49. *New York Review of Books,* April 4, 1996, pp. 68–69.

50. For a good example of why it is sometimes useful to defer judgment when you read a polemical paper, you might wish to examine Farley and Geison's 1974 paper on the Pasteur-Pouchet feud; then Roll-Hansen's 1979 critique of that paper; then Geison's retake of the feud in chapter 5 of his 1995

book; and, finally, in Geison's note on pp. 321–322, his own explanation of why his later description of the feud is different from his earlier paper's.

51. Monaghan, 1995, p. A11.

6. Golgi versus Ramón y Cajal

1. Ramón y Cajal, 1937, p. 564. This portion was probably written in 1907.
2. Ibid., p. 505.
3. Ibid., p. 305.
4. Cannon, 1949, p. 134.
5. Ramón y Cajal, 1937, p. 322.
6. Ibid., p. 289.
7. Quoted in Addison, 1930, p. 182.
8. Ibid.
9. Ramón y Cajal, 1937, p. 553.
10. Ibid., p. 560.
11. Ibid., p. 561.
12. Clifford, 1974.
13. In Restak, 1995, p. 39.
14. Loewy, 1971, p. 9.
15. Quoted in ibid., p. 17.
16. Ramón y Cajal, 1937, p. 3.
17. Ibid., p. 14.
18. Ibid., p. 510.
19. Ibid., p. 252.
20. Ibid., p. 304.
21. Ibid., p. 534.
22. Rothman, 1985, p. 76.
23. Rothman and Orci, 1996, pp. 70–75; Ross, 1989, p. 28.
24. Greenfield, 1997, pp. 79, 80.
25. "Introduction and Perspectives," in Santini, 1975, p. xiii.
26. Ibid.
27. Ibid., p. xvi.
28. Ibid., p. xvii.
29. J. Szentágothai, "What the 'Reazione Nera' Has Given to Us," in Santini, 1975, p. 3.
30. Addison, 1930, p. 178.
31. Bergey, 1995, p. 1088.
32. Now more commonly shortened to *neuron*.
33. Ramón y Cajal, 1937, p. 586. Russell A. Johnson, archivist at the

UCLA Biomedical Library, sees the omission by the French author (Jacques [?] Poirier) as merely reflecting his ignorance of Spanish publications, rather than a decline in Cajal's reputation. It seems to me the result was equally galling to Cajal, and equally indicative of his reputation at the time.

34. This was the first time that the Nobel Prize in physiology or medicine was shared.

35. I thank Russell A. Johnson for this suggestion.

36. Greenfield, 1997, pp. 82, 83.

7. Freud versus Moll, Breuer, Jung, and Many Others

1. Thomas Harris, *The Silence of the Lambs* (New York: St. Martin's Press, 1988), p. 211.

2. Freud, 1952, pp. 81–82.

3. For a fictional account of how a turn-of-the-century psychologist—often called an alienist at the time—might have used such insights to track down a dangerous murderer, see *The Alienist,* by Caleb Carr (New York: Random House, 1994).

4. Quoted in Philip Rieff, "The Ethic of Honesty," in Bloom, 1985, p. 18.

5. Ibid.

6. For example, Hippolyte Bernheim, in the mid-1880s, on the idea that for both neurotics and criminals, the will may not be a free agent.

7. Bloom, 1985, pp. 1, 2.

8. The Freud collection at the U.S. Library of Congress alone consists of more than 50,000 items in diverse formats (James H. Billington, Librarian of Congress, cited in Roth, 1998, p. x.)

9. Brief remarks at a commemorative ceremony held at the New York Academy of Sciences, November 4, 1999.

10. Letter to Dr. van Eeden, a Dutch psychopathologist, December 28, 1914, in Jones, 1955, vol. 2, p. 368.

11. A wide-ranging concept that has little to do with specific sex acts, but rather with castration anxiety, Oedipal cravings, penis envy, and other such considerations. In his autobiography, Freud refers to "the detaching of sexuality from the genitals . . ." (p. 71).

12. Bettelheim, 1990, p. 4.

13. Bloom, 1985, p. 2.

14. Jones, 1955, vol. 2, p. 12.

15. Though the publication date is 1909, it was apparently available by late 1908.

16. By 1908, Freud had already found confirmation of his approach via direct observation of children (Freud, 1952, p. 72).

17. It was not uncommon at the time for speakers to use the third person when referring to their own work. The quote is drawn from the minutes of the meeting. Sulloway, 1979, p. 471.

18. Jones, 1955, vol. 2, p. 114.

19. In Roazen, 1975, p. 194.

20. In Sulloway, 1979, pp. 470–471.

21. Ibid., p. 469.

22. Clark, 1980, p. 56.

23. Freud, 1952, p. 28.

24. Cathartic in the sense of some sort of emotional release.

25. Anna O., or Bertha Pappenheim, coined this term because she found that talking about her troubles and feelings seemed to make her feel better.

26. In Brill, 1966, p. 10.

27. Jones, 1953, vol. 1, p. 337.

28. Freud, 1952, p. 34.

29. In the myth, Oedipus kills his father and marries his mother; in the complex, the young child desires a vaguely sexual union with the parent of the opposite sex and feels a hostile rivalry with the other parent. The idea has fallen out of favor among most modern analysts, but remains a powerful image.

30. Sulloway, 1979, p. 432.

31. McGuire, 1974, p. 230.

32. All quotes from Schultz, 1990, p. 224.

33. Jones, 1955, vol. 2, p. 66.

34. Recent biomedical advances, such as imaging of the brain's activities (blood flow, electrical synapses at work, etc.) have led to renewed hopes that these phenomena can be correlated with mental activities, both normal and abnormal. In this way, perhaps the connection between the brain and the mind can be elucidated. See, e.g., Mlot, 1998, and Niehoff, 1999; also see Restak, 1995, and Greenfield, 1997, both in the bibliography for chapter 6.

35. Thomas Szasz began his attack in the 1970s. Other attackers include R. D. Laing, Peter J. Swales, Erving Goffman, and, oddly, the novelist Ken Kesey. His 1962 novel *One Flew Over the Cuckoo's Nest* had an enormous impact on the public, especially after it was made into a blockbuster movie in 1975. See, e.g., Porter, 1998, pp. 522, 523; also Torrey, 1974.

36. Shorter, 1997, p. 325.

37. Cioffi, 1970, pp. 471–499.

38. Victor Tausk was one of Freud's early followers. Sulloway carries this idea even further, claiming that Freud was responsible for six suicides among his followers (Sulloway, 1979, p. 482).

39. Malcolm, 1984, p. 114. Freudians, of course, take issue with such

claims. See, e.g., Malcolm on Eissler (1984, p. 8). For a brief summary of Swales's position, in his own words, see Malcolm, 1984, pp. 134, 135.

40. Jones, 1955, vol. 2, pp. 120–122 and, especially, p. 404.

41. Roazen, 1975, p. 181.

42. Lomas, 1996, p. 95.

43. Jones, 1953, vol. 1, pp. 292–293.

44. Freud, 1952, pp. 14, 15.

45. July 23, 1908, in Jones, 1955, vol. 2, pp. 49–50.

46. Ibid., p. 398.

47. Roazen, 1975, p. 195.

48. Ibid., p. 196.

49. Ibid., p. 189.

50. Freud, 1952, pp. 26, 27.

51. Jones, 1953, vol. 1, p. 330.

52. Sulloway, 1979, p. 464.

53. Masson, 1985, p. 185.

54. Jones, 1953, vol. 1, p. 292.

55. Freud, 1952, p. 62.

56. Bragg, 1998, p. 227.

57. Jones, 1955, vol. 2, pp. 108, 109.

58. Bettelheim, 1990, p. 12.

59. Sulloway, 1979, p. 468.

60. Ibid., p. 421. For an update on Sulloway's thinking, see his articles, "The Rhythm Method" and "Exemplary Botches," in Crews, 1998, pp. 54–68 and 174–185.

61. Richard Ofshe, coauthor of *Making Monsters: False Memories, Psychotherapy, and Sexual Hysteria* (New York: Scribner's, 1994), in blurb on back cover of Crews, 1998.

62. Crews, 1998, p. x.

63. See, e.g., Dimen, 1998, pp. 207–220.

64. For example, several articles by Crews, starting in 1994, and another, "Freud Under Analysis," by Colin McGinn (November 4, 1999, p. 20).

65. Grünbaum, 1998, p. 185.

66. Lear, 1998, p. 32.

67. Gay, 1999, p. 66.

68. Kiell, 1988, p. 3.

69. Ibid.

70. Ibid., p. 13.

71. Ibid., p. 15.

8. Sabin versus Salk

1. Carter, 1966, p. 3.
2. Ibid., p. 6.
3. Nuland, 1995, p. 50.
4. Sheed, 1999, p. 170.
5. *New York Times,* April 13, 1955, p. 23.
6. Oldstone, 1998, pp. 94, 95.
7. One of the first microscopic studies of tissue from polio patients was done around 1870 by Jean-Martin Charcot, with whom Freud studied later on (though not in neurology; see chapter 7). Charcot noted shrinkage in the region of the spinal cord that contains the large motor neurons that control the limbs. This and other details in Oldstone, 1998, pp. 93–95.
8. Hans Zinsser, famous bacteriologist and author of the 1935 book *Rats, Lice, and History,* maintained that epidemics were more efficient than any military strategy at bringing down armies (pp. 9–10).
9. Rogers, 1992, p. 4.
10. For a fuller picture of this process, see Klein, 1972, pp. 38–41, 50–55.
11. Carter, 1966, p. 81.
12. Oldstone, 1998, p. 107.
13. Carter, 1966, p. 82.
14. The main reason room was available was the development of antibiotics, which helped empty a large area in the university hospital that had once been packed with victims of bacteria-based infectious diseases.
15. Friedman and Friedland, 1998, p. 152.
16. This, ironically, includes the work of Ross Granville Harrison, who, in 1907, developed the first real tissue culture method. He was nominated twice for a Nobel Prize, but was passed over both times. See, e.g., ibid., pp. 133–152.
17. *A Paralyzing Fear: The Story of Polio in America,* videorecording, distributed by PBS Home Video, 1998.
18. Klein, 1972, p. 71.
19. See, e.g., Richard Thayer Goldberg, *The Making of Franklin D. Roosevelt: Triumph Over Disability* (Cambridge, Mass.: Abt Books), 1981.
20. Cimons, 1983, p. A11.
21. Carter, 1966, p. 82.
22. Gould, 1995, p. 129.
23. See, e.g., ibid., pp. 132–152, for more details on what turned into a harrowing experience for all.
24. Salk, 1953, pp. 1081–1098.
25. Klein, 1972, p. 83.

26. Smith, 1990, p. 193.

27. Klein, 1972, pp. 105, 106.

28. For more detail on this, see, e.g., Smith, 1990, pp. 359–367.

29. In 1948, Dr. Isabel Morgan at Johns Hopkins had successfully immunized monkeys against polio using a killed-virus vaccine.

30. Nuland, 1995, pp. 49, 50.

31. Sheed, 1999, p. 170.

32. "Age of Salk's Aides Averages under 40," *New York Times,* April 13, 1955, p. 23.

33. Nuland, 1995, p. 50.

34. Ibid.

35. Cimons, 1983, p. A11.

36. Ibid.

37. Grady, 1999, pp. F1, F5.

38. In Sheed, 1999, p. 170.

39. Smith, 1990, p. 376.

40. Podolsky, 1998, p. 269.

9. Franklin versus Wilkins

1. Watson and Crick, 1953, pp. 737–738.

2. There is currently a major effort in the biomedical world to apply gene therapy to the treatment and perhaps cure of hemophilia. See, e.g., Garber, 2000, pp. 58–60, 62, 64.

3. See, e.g., "Winning the Cancer War," by Craig Horowitz, *New York,* February 7, 2000, pp. 26–33.

4. Hollar, in Magill, 1991, p. 839.

5. Sayre, 1975, p. 40.

6. Ibid., pp. 39, 41.

7. Ibid., pp. 40, 41.

8. Ibid., p. 32.

9. Ibid., p. 26.

10. Ibid., p. 28.

11. Judson, 1979.

12. Judson, 1986, pp. 78–79.

13. Friedman and Friedland, 1998, pp. 206–207.

14. Letter from Randall to Franklin, December 4, 1950, in Olby, 1974, p. 346.

15. Crick, 1988, p. 69.

16. Olby, 1974, p. 344.

17. Chargaff, 1978, p. 101.

18. Judson, 1979, p. 143.

19. Ibid., p. 143.

20. Crick, 1988, p. 66.

21. Lwoff, 1968, p. 134.

22. Curtis, 1993, p. 256.

23. Perutz, 1969, p. 1537.

24. Today the word "model" can refer to a computer model. In the 1950s the word referred to an actual three-dimensional representation of a molecular structure, generally fashioned from cardboard, wood, or metal.

25. Watson, 1968, pp. 164–167.

26. Judson, 1979, p. 159.

27. Judson, 1986, p. 81.

28. Watson, 1968, p. 167.

29. Sayre, 1975, pp. 151–152.

30. Judson, 1979, p. 144.

31. Crick, 1988, p. 60.

32. Ibid., p. 69.

33. Ibid., p. 68.

34. Ibid., p. 71.

35. Watson, 1968, p. 205.

36. Crick, 1988, p. 76.

37. Curtis, 1993, p. 257.

38. Sayre, 1975, p. 171.

39. Watson, 1968, pp. 68–69.

40. Sayre, 1975, p. 172.

41. Olby, 1970, p. 963.

42. Watson, 1968, p. 225.

43. Ibid., pp. 225, 226.

44. Sayre, 1975, p. 143.

45. Bate and Gaskin, 1983, p. 12.

46. Sayre, 1975, p. 196.

47. Crick, 1988, pp. 68–69.

48. Ibid., p. 69.

49. Judson, 1986, p. 78.

50. Ibid., p. 80.

51. Ibid.

52. See, for example, Kass-Simon and Farnes, 1990, pp. 335–348.

53. Judson, 1986, p. 80.

54. Ibid.

55. Crick, 1988, p. 69.

56. Pauling, 1974, p. 771.

57. Sayre, 1975, p. 220.

58. Friedman and Friedland, 1998, p. 225.

59. See, e.g., Bernstein, 1978, p. 153.

60. Crick, 1988, p. 76.

10. Gallo versus Montagnier

1. Altman, 1999, p. F6.

2. Gallo, 1991, p. 7.

3. Personal interview, March 22, 2000.

4. *Morbidity and Mortality Weekly Report,* June 5, 1981, vol. 30, no. 1, pp. 250, 251.

5. Schoub, 1994, p. 2, and *Discovery,* the newsletter of the Institute of Human Virology, Spring 1999, p. 2.

6. Schoub, 1994, p. 3.

7. Gallo, 1991, p. 46.

8. Ibid., p. 68.

9. Montagnier, 2000, p. 121.

10. Sometimes written HTLV-I and HTLV-II.

11. Shilts, 1988, p. 264.

12. Montagnier, 2000, pp. 60, 61.

13. Gallo, 1991, pp. 157, 158.

14. Ibid., pp. 163, 164.

15. Ibid., p. 165.

16. Ibid., p. 170.

17. Ibid., pp. 168, 169.

18. Ibid., p. 169.

19. Ibid., p. 170.

20. For a strongly anti-Gallo view, see Duesberg, 1996, pp. 159–167. Shilts's book, too, puts Gallo in a negative light.

21. Gallo, 1991, pp. 175, 176.

22. Ibid., p. 176.

23. Ibid., p. 180.

24. Montagnier, 2000, p. 73.

25. Ibid., p. 69.

26. Both quotes from personal communication, May 1, 2000.

27. Shilts suggests that earlier negotiations between representatives of the Pasteur Institute, who felt that credit for finding the virus belonged to them, and with and between the NCI and the CDC, who were themselves in some sort of competition, came to naught; he also says that announcing the development as an American one was Gallo's idea, and that leaks to the press

of both claims led to a quick mounting of the NCI/NIH-sponsored press conference. (Shilts, 1988, pp. 444, 445.)

28. Personal interview with Gallo, March 28, 2000, and Shilts, 1988, p. 451.

29. Even today, after such a test has been developed, the World Health Organization estimates that 5 to 10 percent of HIV infections occur through transmission of infected blood or blood products. (*New York Times,* April 10, 2000, p. A18.)

30. Editorial, "A Viral Competition over AIDS," *New York Times,* April 26, 1984, p. 22. Also see Shilts, 1988, p. 451.

31. Shilts, 1988, p. 452.

32. Gallo, 1991, p. 170.

33. All Shilts, 1988, pp. 528, 529.

34. Oldstone, 1998, p. 155.

35. Montagnier, 2000, p. 82.

36. Also known as Mikulas, a virologist who had been investigated and cleared along with Dr. Gallo.

37. Montagnier, 2000, p. 82.

38. Jasanoff, 1997, p. 476.

39. Horton, 1996, p. 16.

40. Gallo, 1991, pp. 8, 9.

41. See, e.g., Hamilton, November 15, 1991, p. 944, and Cohen and Marshall, 1994, p. 313. In the latter article, the authors state: "Perhaps the only remaining player who could reignite the Gallo affair is Representative John Dingell (D-MI), whose probe into Gallo's lab has not yet been concluded."

42. Cohen, July 1, 1994, p. 23.

43. Kaiser, 1997, pp. 920, 921.

44. Cohen, "Report Card," November 15, 1991, pp. 948, 949.

45. P. 342.

46. For a more complete discussion, see, e.g., Cohen, July 1, 1994, pp. 23–25.

47. Personal communication, May 1, 2000.

48. Personal interview, March 22, 2000.

49. Personal communication, May 2, 2000.

Epilogue

1. Personal communication, May 2, 2000.

BIBLIOGRAPHY

General Background

Abse, Dannie. *Medicine on Trial.* New York: Crown Publishers, 1969.

Bolles, Edmund Blair, ed. *Galileo's Commandment: An Anthology of Great Science Writing.* New York: W. H. Freeman, 1997.

Boynton, Holmes, ed. *The Beginnings of Modern Science: Scientific Writings of the 17th and 18th Centuries.* New York: Walter J. Black, 1948.

Clendening, Logan, ed. *Source Book of Medical History.* 1942. Reprint, New York: Dover, 1960.

Dampier-Whetham, William Cecil Dampier. *A History of Science and Its Relations with Philosophy and Religion.* New York: Macmillan, 1931.

Dascal, Marcelo. "The Study of Controversies and the Theory and History of Science." *Science in Context,* Summer 1998, vol. 11, no. 2, pp. 147–154.

Eliot, Charles W., ed. *Scientific Papers: Physiology, Medicine, Surgery, Geology.* Harvard Classics, vol. 38, 1897. Reprint, New York: P. F. Collier and Son, 1910.

Freudenthal, Gideon. "Controversy." *Science in Context,* Summer 1998, vol. 11, no. 2, pp. 155–160.

Friedman, Meyer, and Gerald W. Friedland. *Medicine's 10 Greatest Discoveries.* New Haven, Conn.: Yale University Press, 1998.

Gillispie, Charles C., ed. *Dictionary of Scientific Biography.* 16 vols. New York: Scribner, 1970–1980.

Gross, Paul R., and Norman Levitt. *Higher Superstition: The Academic Left and Its Quarrels with Science.* Baltimore: Johns Hopkins University Press, 1994.

Haggard, Howard W. *Devils, Drugs, and Doctors: The Story of the Science of Healing from Medicine-Man to Doctor.* New York: Harper and Brothers, 1929.

Hall, A. Rupert. "Medicine and the Royal Society." In Debus, 1974, pp. 421–452.

Hellman, Hal. *Great Feuds in Science: Ten of the Liveliest Disputes Ever.* New York: John Wiley and Sons, 1998.

Inglis, Brian. *A History of Medicine.* Cleveland: World Publishing, 1965.

Isaacs, Ronald H., *Judaism, Medicine, and Healing*. Northvale, N.J.: Jason Aronson, 1998.

Kelsey, Morton T. *Healing and Christianity: In Ancient Thought and Modern Times*. 1973. Reprint, New York: Harper and Row, 1976.

Magill, Frank N., ed. *The Nobel Prize Winners: Physiology or Medicine*. Vol. 1, *1901–1944*, vol. 2, *1944–1969*. Pasadena: Salem Press, 1991.

Magner, Lois N. *A History of Medicine*. New York: Marcel Dekker, 1992.

Nuland, Sherwin B. *Doctors: The Biography of Medicine*. New York: Alfred A. Knopf, 1988.

Oldstone, Michael B. A. *Viruses, Plagues, and History*. New York: Oxford University Press, 1998. (For the basics of virology and immunology, see chapters 1, 2, and 3.)

Porter, Roy. *The Greatest Benefit to Mankind: A Medical History of Humanity*. New York: W. W. Norton, 1998.

——, ed. *Cambridge Illustrated History of Medicine*. Cambridge, Eng.: Cambridge University Press, 1996.

Robinson, Victor. *Pathfinders in Medicine*. Exp. ed. 1912. Reprint, New York: Medical Life Press, 1929.

Sigerist, Dr. Henry E. *The Great Doctors*. 1933. Reprint, New York: Dover Publications, 1971.

Silver, Brian L. *The Ascent of Science*. New York: Oxford University Press, 1998.

Simmons, John. *The Scientific 100: A Ranking of the Most Influential Scientists, Past and Present*. Secaucus, N.J.: Citadel Press, 1996.

Taton, René. *The Beginnings of Modern Science*. 1958. Reprint, New York: Basic Books, 1964.

Thomas, Lewis. *The Medusa and the Snail: More Notes of a Biology Watcher*. New York: Bantam, 1980.

Weisse, Allen B. *Medical Odysseys: The Different and Sometimes Unexpected Pathways to Twentieth-Century Medical Discoveries*. New Brunswick, N.J.: Rutgers University Press, 1991. (See the chapters titled "Sparks, Life-Giving Electricity," pp. 135–157, and "Polio, the Not-So-Twentieth-Century Disease," pp. 158–185.)

Weissman, Gerald. *They All Laughed at Christopher Columbus*. New York: Times Books, 1987.

White, Andrew Dickson. *A History of the Warfare of Science with Theology in Christendom*. New York: Free Press, 1965.

Wightman, W. P. D. *The Growth of Scientific Ideas*. New Haven, Conn.: Yale University Press, 1953.

Williams, Guy. *The Age of Agony: The Art of Healing, 1700–1800*. Chicago: Academy Chicago Publishers, 1986.

Introduction

Auenbrugger, Leopold. *On Percussion of the Chest* (1761). In J. N. Corvisart, *An Essay on the Organic Diseases and Lesions of the Heart and Great Vessels.* Trans J. Gates, Boston n.p., 1812. Quoted in Nuland, 1988, p. 204.

Debus, Allen G. "Harvey and Fludd: The Irrational Factor in the Rational Science of the Seventeenth Century." *Journal of the History of Biology,* Spring 1970, vol. 3, no. 1, pp. 81–105.

1. Harvey versus Primrose, Riolan, and the Anatomists

Bylebyl, Jerome, J. "Harvey, William." In Gillispie, 1972, vol. 6, pp. 150–162.

———. "The Medical Side of Harvey's Discovery: The Normal and the Abnormal." In Bylebyl, 1979, pp. 28–102.

———, ed. *William Harvey and His Age: The Professional and Social Context of the Discovery of the Circulation.* Baltimore: Johns Hopkins University Press, 1979.

Chauvois, Louis. *William Harvey: His Life and Times, His Discoveries, His Methods.* New York: Philosophical Library, 1957.

Debus, Allen G. "Harvey and Fludd: The Irrational Factor in the Rational Science of the Seventeenth Century." *Journal of the History of Biology,* Spring 1970, vol. 3, no. 1, pp. 81–105.

———, ed. *Science, Medicine and Society in the Renaissance.* Essays to honor Walter Pagel, vol. 2. New York: Science History Publications, 1972.

———. *Medicine in Seventeenth Century England.* A symposium held at UCLA in honor of C. D. O'Malley. Berkeley: University of California Press, 1974.

Dick, Oliver Lawson. *Aubrey's Brief Lives.* 1813. Reprint, Ann Arbor: University of Michigan Press, 1957.

Frank, Robert G., Jr. "The Image of Harvey in Commonwealth and Restoration England." In Bylebyl, 1979, pp. 103–143.

———. *Harvey and the Oxford Physiologists: A Study of Scientific Ideas.* Berkeley: University of California Press, 1980.

Franklin, Kenneth J. *William Harvey, Englishman.* London: MacGibbon and Kee, 1961.

French, Roger. *William Harvey's Natural Philosophy.* Cambridge, Eng.: Cambridge University Press, 1994.

Gorham, Geoffrey. "Mind-Body Dualism and the Harvey-Descartes Controversy." *Journal of the History of Ideas,* 1994, vol. 55, no. 2, pp. 211–234.

Greenblatt, Robert B. *Search the Scriptures: A Physician Examines Medicine in the Bible*. Philadelphia: J. B. Lippincott, 1963.

Hamburger, Jean. *The Diary of William Harvey: The Imaginary Journal of the Physician Who Revolutionized Medicine*. New Brunswick, N.J.: Rutgers University Press, 1992.

Harvey, William. *On the Motion of the Heart and Blood in Animals*. Trans. by Robert Willis. 1847. Reprint, Buffalo, N.Y.: Prometheus Books, 1993. (The U.S. edition was originally published by P. F. Collier and Son, New York, 1910. This translation is also available in Harvard Classics, vol. 38, pp. 59–139.)

Hobbes, Thomas. *Leviathan*. 1651. Reprint, New York: Penguin, 1986.

Keynes, Geoffrey. *The Life of William Harvey*. Oxford: Clarendon Press, 1966.

Majno, Guido. *The Healing Hand: Man and Wound in the Ancient World*. 1975. Reprint, Cambridge, Mass.: Harvard University Press, 1991.

Mani, Nikolaus. "Jean Riolan II (1580–1657) and Medical Research." *Bulletin of the History of Medicine*, 1968, vol. 42, pp. 121–144.

Packard, F. R. *Guy Patin and the Medical Profession in Paris in the Seventeenth Century*. London: Oxford University Press, 1924.

Pagel, Walter. *William Harvey's Biological Ideas: Selected Aspects and Historical Background*. New York: S. Karger, 1967.

Rattansi, P. M. "The Helmontian-Galenist Controversy in Restoration England," *Ambix*, February 1964, vol. 12, no. 1, pp. 1–23.

Silvette, Herbert. *The Doctor on the Stage: Medicine and Medical Men in Seventeenth-Century England*. Knoxville: University of Tennessee Press, 1967.

Toellner, Richard. "The Controversy between Descartes and Harvey Regarding the Nature of Cardiac Motions." In Debus, 1972, pp. 73–89.

Trent, J. C. "Five Letters of Marcus Aurelius Severinus." *Bulletin of the History of Medicine*, 1944, vol. 15, pp. 306–323.

Webster, Charles. "William Harvey and the Crisis of Medicine in Jacobean England." In Bylebyl, 1979, pp. 1–27.

Weil, E. "The Echo of Harvey's *De Motu Cordis* (1628)." *Journal of the History of Medicine and Allied Sciences*, April 1957, pp. 167–174. (Also found in slightly different form in Keynes, 1966, pp. 447–455.)

Weiss, David. *Physician Extraordinary: A Novel of the Life and Times of William Harvey*. New York: Delacorte, 1975.

Whitteridge, Gweneth. *William Harvey and the Circulation of the Blood*. London: Macdonald, 1971. (See especially the material on Riolan, pp. 175–182.)

Yount, Lisa. *William Harvey: Discoverer of How Blood Circulates*. Great Minds of Science Series. Springfield, N.J.: Enslow Publishers, 1994.

2. Galvani versus Volta

Arago, Dominique. "Eulogy on Alexander Volta." Smithsonian Institution Annual Report. Washington, D.C.: 1875, pp. 115–141.

Brown, Theodore M. "Galvani, Luigi." In Gillispie, 1972, vol. 5, pp. 267–269.

Buckley, Arabella B. *A Short History of Natural Science and of the Progress of Discovery from the Time of the Greeks to the Present Day: For the Use of Schools and Young Persons.* New York: D. Appleton, 1881. (See especially pp. 259–264.)

de Santillana, Giorgio. "Alessandro Volta." *Scientific American,* December 1965, pp. 82–91.

Dibner, Bern. *Galvani-Volta.* Norwalk, Conn.: Burndy Library, 1952.

——. *Alessandro Volta and the Electric Battery.* New York: Franklin Watts, 1964.

——. *Luigi Galvani.* Norwalk, Conn.: Burndy Library, 1971.

Fulton, J. F., and H. Cushing. "A Bibliographical Study of the Galvani and Aldini Writings on Animal Electricity." *Annals of Science,* 1936, vol. 1, pp. 239–268.

Galvani, Luigi. *Commentary on the Effect of Electricity on Muscular Motion.* Trans. R. M. Green. Cambridge, Mass.: Licht, 1953.

——. *Commentary on the Effects of Electricity on Muscular Motion.* 1791. Trans. Margaret Glover Foley. Introd. by I. Bernard Cohen. With facsimile of 1791 ed. Norwalk, Conn.: Burndy Library, 1954.

Geddes, L. A., and H. E. Hoff. "The Discovery of Bioelectricity and Current Electricity." *IEEE Spectrum,* December 1971, pp. 38–46.

Heilbron, J. L. "Volta, Alessandro Giuseppe Antonio Anastasio." In Gillispie, 1976, vol. 14, pp. 69–81.

——. *Electricity in the Seventeenth and Eighteenth Centuries.* Berkeley: University of California Press, 1979. (See especially pp. 490–495.)

Hoff, H. E. "Galvani and the Pre-Galvanian Electrophysiologists." *Annals of Science,* April 15, 1936, vol. 1, pp. 157–172.

Kellaway, Peter. "The Part Played by Electric Fish in the Early History of Bioelectricity and Electrotherapy." *Bulletin of the History of Medicine,* 1946, vol. 20, pp. 112–137.

Klopfer, Leo E. *Frogs and Batteries.* History of Science Cases for High Schools. Middletown, Conn.: Wesleyan University, 1960.

Lenard, Philipp. *Great Men of Science.* New York: Macmillan, 1933, pp. 158–170.

Mauro, Alexander. "The Role of the Voltaic Pile in the Galvani-Volta Controversy Concerning Animal vs Metallic Electricity." *Journal of the History of Medicine,* April 1969, vol. 24, pp. 140–150.

Pera, Marcello. *The Ambiguous Frog: The Galvani-Volta Controversy on*

Animal Electricity. Trans. Joseph Mandelbaum. 1986. Reprint, Princeton, N.J.: Princeton University Press, 1992.

Trumpler, Maria. Review of Pera, 1992. In *Isis,* 1994, vol. 85, no. 4, pp. 701, 702.

Volta, Alexander. "Account of Some Discoveries Made by Mr. Galvani, of Bologna; with Experiments and Observations on Them. In two Letters from Mr. Alexander Volta, F. R. S., Professor of Natural Philosophy in the University of Pavia, to Mr. Tiberius Cavallo, F. R. S., p. 10. From the French." *Philosophical Transactions,* 1793, vol. 83, pp. 285–291.

Walker, W. Cameron. "The Detection and Estimation of Electric Charges in the Eighteenth Century." *Annals of Science,* 1936, vol. 1, pp. 66–100.

Walsh, John. "Experiments and Observations of the Torpedo." *Philosophical Transactions,* 1773, vol. 63, pp. 461–477.

3. Semmelweis versus the Viennese Medical Establishment

Carter, K. Codell. "Ignaz Semmelweis, Carl Mayrhofer, and the Rise of Germ Theory." *Medical History,* 1985, vol. 29, pp. 33–53.

Carter, K. Codell, Scott Abbott, and James L. Siebach."Five Documents Relating to the Final Illness and Death of Ignaz Semmelweis." *Bulletin of the History of Medicine,* 1995, vol. 69, pp. 255–270.

Carter, K. Codell, and Barbara R. Carter. *Childbed Fever: A Scientific Biography of Ignaz Semmelweis.* Westport, Conn.: Greenwood Press, 1994.

Carter, K. Codell, and George S. Tate. "The Earliest-Known Account of Semmelweis's Initiation of Disinfection at Vienna's Allgemeines Krankenhaus." *Bulletin of the History of Medicine,* 1991, vol. 65, pp. 252–257.

Haggard, 1929, chapter 4, "A Gentleman with Clean Hands May Carry the Disease," pp. 66–89.

Norman, Elizabeth. "For Want of Soap and Water." *New York Times,* March 27, 2000, p. A21.

Nuland, Sherwin B. "The Enigma of Semmelweis—an Interpretation." *Journal of the History of Medicine,* July 1979, vol. 34, pp. 255–272.

Risse, Guenter B. "Semmelweis, Ignaz Philipp." In Gillispie, 1975, vol. 12, pp. 294–297.

Robinson, 1929, chapter 25, "Ignaz Philipp Semmelweis," pp. 624–647.

Semmelweis, I. P. *The Etiology, the Concept and the Prophylaxis of Childbed Fever.* Trans. F. P. Murphy. Birmingham: Classics of Medicine, 1981. (Includes *The Open Letters,* trans. S. B. Nuland and F. A. Gyorgyey. Selections can be found in various collections; e.g., Clendening, 1960, pp. 606–610.)

——. *The Etiology, the Concept and the Prophylaxis of Childbed Fever.* Trans. K. Codell Carter. Madison: University of Wisconsin Press, 1983.

Sigerist, 1941, chapter 41, "Ignaz Philipp Semmelweis," pp. 354–359.

Slaughter, Frank G. *Immortal Magyar: Semmelweis, Conquerer of Childbed Fever.* New York: Henry Schuman, 1950.

Thompson, Morton. *The Cry and the Covenant.* Garden City, N.Y.: Garden City Books, 1949.

Wangesteen, O. H. "Nineteenth Century Wound Management of the Parturient Uterus and Compound Fracture: The Semmelweis-Lister Priority Controversy." *Bulletin of the New York Academy of Medicine,* August 1970, vol. 46, no. 8, pp. 565–596.

Weissman, Gerald. "Puerperal Priority." *Lancet,* 1997, vol. 349, pp. 122–125.

Yoffe, Emily. "Doctors Are Reminded, 'Wash Up!' " *New York Times,* November 9, 1999, pp. C1, C9.

4. Bernard versus Chemists, Physicians, and Antivivisectionists

Aldhous, Peter, et al. "Let the People Speak: Animal Experiments, Where Do You Draw the Line?" (Opinion poll, plus comments.) *New Scientist,* May 22, 1999, pp. 26–31, 60, 61.

Bateson, Patrick. "Do Animals Feel Pain?" *New Scientist,* April 25, 1992, pp. 30, 32, 33.

Barnard, Neal D., and Stephen R. Kaufman. "Animal Research Is Wasteful and Misleading." *Scientific American,* February 1997, vol. 276, no. 2, pp. 80–82.

Bernard, Claude. "On the Functions of the Brain." *Popular Science Monthly,* November 1872, vol. 2, pp. 64–74.

———. "The Definition of Life." *Popular Science Monthly–Supplement,* 1878, vols. 7–12, pp. 511–524.

———. *An Introduction to the Study of Experimental Medicine.* 1927. Reprint, New York: Dover Publications, 1957.

Blum, Deborah. *The Monkey Wars.* New York: Oxford University Press, 1994. (About animal rights and animal testing.)

Botting, Jack H., and Adrian R. Morrison. "Animal Research Is Vital to Medicine." *Scientific American,* February 1997, vol. 276, no. 2, pp. 83–85.

Bowden, Douglas M. "Animals in the Laboratory." Letter to the editor. *Science,* May 14, 1982, pp. 682, 684.

Coile, D. Caroline, and Neal E. Miller. "How Radical Animal Activists Try to Mislead Humane People." *American Psychologist,* June 1984, vol. 39, p. 701.

Darwin, Charles. *The Autobiography of Charles Darwin and Selected Letters.* Ed. Francis Darwin. 1892. Reprint, New York: Dover, 1958, pp. 303–308.

Editor. "Philanthropic Fanaticism against Science." *Popular Science Monthly,* January 1877, vol. 10, pp. 363–366.

——. "Sketch of Claude Bernard." *Popular Science Monthly,* October 1878, vol. 13, pp. 742–744.

Foster, Michael. *Claude Bernard.* New York: Longmans, Green, 1899.

Fox, Michael Allen. *The Case for Animal Experimentation: An Evolutionary and Ethical Perspective.* Berkeley: University of California Press, 1986.

Galant, Mavis. "Fairy Tales and Other Cruelties." Review of *The Great Cat Massacre and Other Episodes in French Cultural History,* by Robert Darnton (New York: Basic Books, 1984). In *New York Times Book Review,* February 12, 1984, p. 12.

Grmek, M. D. "Bernard, Claude." In Gillispie, 1970, vol. 2, pp. 24–34.

Guillermo, Kathy Snow. *Monkey Business.* Washington, D.C.: National Press Books, 1993. (The story of an animal abuse case, and of the history of PETA [People for the Ethical Treatment of Animals], told by a PETA official.)

Hampson, Judith. "Animal Welfare—a Century of Conflict." *New Scientist,* October 25, 1979, pp. 280–282.

Hilts, Philip J. "A History: When the State Uses People as Guinea Pigs." *New York Times,* July 13, 1999, p. F2.

Hoff, H. E. Review of Claude Bernard's *An Introduction to the Study of Experimental Medicine. Bulletin of the History of Medicine,* 1962, vol. 36, pp. 177–181.

Hoge, Warren. "British Researchers on Animal Rights Death List." *New York Times,* January 10, 1999, p. A10.

Holmes, Frederic L. *Claude Bernard and Animal Chemistry: The Emergence of a Scientist.* Cambridge, Mass.: Harvard University Press, 1974.

Kolata, Gina. "Tough Tactics in One Battle over Animals in the Lab." *New York Times,* March 24, 1998, pp. E1, E6.

Masci, David. "Fighting over Animal Rights." *CQ Researcher,* August 2, 1996, vol. 6, no. 29, pp. 673–696.

Mukerjee, Madhusree. "Trends in Animal Research." *Scientific American,* February 1997, vol. 276, no. 2, pp. 86–93.

Nozick, Robert. "About Mammals and People." Essay review of Tom Regan's *The Case for Animal Rights. New York Times Book Review,* November 27, 1983, pp. 11, 29–30.

Olmsted, J. M. D. "The Contemplative Works of Claude Bernard." *Bulletin of the Institute for the History of Medicine,* 1935, vol. 3, no. 5, pp. 335–354.

——. *Claude Bernard, Physiologist.* New York: Harper and Brothers, 1938.

Olmsted, J. M. D., and E. Harris Olmsted. *Claude Bernard and the Experimental Method in Science.* New York: Henry Schuman, 1952.

Parascandola, Mark. "Epidemiology Second-Rate Science." *Public Health Reports,* July–August 1998, vol. 113, no. 4, p. 312.

Regan, Tom. *The Case for Animal Rights.* Berkeley: University of California Press, 1983.

Riese, W. "Claude Bernard in the Light of Modern Science (Essence, Revision and New Foundations of the Experimental Method)." *Bulletin of the History of Medicine,* October 1943, vol. 14, no. 3, pp. 281–294.

Robin, Eugene Debs. *Claude Bernard and the Internal Environment.* New York: Marcel Dekker, 1979.

Robinson, 1929, chapter 23, "Claude Bernard," pp. 580–610.

Rowan, Andrew N. "The Benefits and Ethics of Animal Research." *Scientific American,* February 1997, vol. 276, no. 2, p. 79.

Schillier, Joseph. "Claude Bernard and Vivisection." *Journal of the History of Medicine and Allied Sciences,* July 1967, vol. 22, pp. 246–260.

Sullivan, Mark D. "Reconsidering the Wisdom of the Body: An Epistemological Critique of Claude Bernard's Concept of the Internal Environment." *Journal of Medicine and Philosophy,* 1990, vol. 15, pp. 493–514.

Tarshis, Jerome. *Claude Bernard: Father of Experimental Medicine.* New York: Dial, 1968.

Tattersall, Robert. "Frederick Pavy (1829–1911) and His Opposition to the Glycogenic Theory of Claude Bernard." *Annals of Science,* July 1997, vol. 54, no. 4, pp. 361–374.

Turney, Jon. *Frankenstein's Footsteps: Science, Genetics and Popular Culture.* New Haven, Conn.: Yale University Press, 1998. (See especially chapter 3, on experimental biology.)

Virtanen, Reino. *Claude Bernard and His Place in the History of Ideas.* Lincoln: University of Nebraska Press, 1960.

Vogt, Carl. "Personal Reminiscences of Deceased Savants." *Popular Science Monthly,* May 1878, vol. 13, pp. 20–25.

Williams, Joy. "The Inhumanity of the Animal People." *Harper's Magazine,* August 1997, pp. 60–67.

Wilson, D. Wright. "Claude Bernard." *Popular Science Monthly,* June 1914, pp. 567–578.

5. Pasteur versus Liebig, Pouchet, and Koch

Debré, Patrice. *Louis Pasteur.* Trans. Elborg Forster. Baltimore: Johns Hopkins Press, 1998.

Debus, Allen G. *The French Paracelsians: The Chemical Challenge to Med-*

ical and Scientific Tradition in Early Modern France. New York: Cambridge University Press, 1991.

de Kruif, Paul. *Microbe Hunters.* New York: Harcourt, Brace, 1926.

——. *The Fight for Life.* New York: Harcourt, Brace, 1938.

Dixon, Bernard. "OK, So Nobody's Perfect." *New Scientist,* January 24, 1995, p. 46. (Review of Geison, 1995.)

Dubos, René. *Louis Pasteur, Free Lance of Science.* 1950. Reprint, New York: Scribner, 1976.

——. *Pasteur and Modern Science.* Rev. ed. 1960. Reprint, Madison, Wis.: Science Tech Publishers, 1988.

Duclaux, Emile. *Pasteur: The History of a Mind.* Philadelphia: W. B. Saunders, 1920.

Farley, John. *The Spontaneous Generation Controversy from Descartes to Oparin.* Baltimore: Johns Hopkins University Press, 1977.

Farley, John, and Gerald L. Geison. "Science, Politics and Spontaneous Generation in Nineteenth-Century France: The Pasteur-Pouchet Debate." *Bulletin of the History of Medicine,* Summer 1974, vol. 48, no. 2, pp. 161–198.

Fee, Elizabeth. "Book Reviews." *New England Journal of Medicine,* September 28, 1995, pp. 884, 885. (Review of Geison, 1995.)

Geison, Gerald L. "Pasteur, Louis." In Gillispie, 1974, vol. 10, pp. 350–417.

——. "Pasteur on Vital versus Chemical Ferments: A Previously Unpublished Paper on the Inversion of Sugar." *Isis,* 1981, vol. 72, no. 263, pp. 425–443.

——. *The Private Science of Louis Pasteur.* Princeton, N.J.: Princeton University Press, 1995.

Grant, Madeleine P. *Louis Pasteur, Fighting Hero of Science.* New York: McGraw-Hill, 1959.

Hendrick, Robert. "Biology, History & Louis Pasteur." *American Biology Teacher,* November–December 1991, vol. 53, no. 8, pp. 467–477.

Metchnikoff, Elie. *The Founders of Modern Medicine.* New York: Walden Publications, 1939. (See especially pp. 9–34.)

Monaghan, Peter. "Separating Fact from Legend." *Chronicle of Higher Education,* September 29, 1995, vol. 42, no. 5, pp. A10–A11.

Pasteur, Louis. "The Physiological Theory of Fermentation." 1879. In Eliot, 1892, pp. 275–363.

——. "Prevention of Rabies: A Method by Which the Development of Rabies after a Bite May Be Prevented." 1885. In Metchnikoff, 1939, pp. 379–387.

Perutz, M. F. "The Pioneer Defended." *New York Review of Books,* December 21, 1995, pp. 54–58. (Review of Geison, 1995. See also responses in the April 4, 1996, issue.)

Porter, Roy. "Lion of the Laboratory: Pasteur's Amazing Achievements Survive the Scrutiny of His Notebooks." *Times Literary Supplement,* June 16, 1995, pp. 3, 4. (Review of Geison, 1995.)

Reynolds, Moira Davison. *How Pasteur Changed History: The Story of Louis Pasteur and the Pasteur Institute.* Bradenton, Fla.: McGuinn and McGuire, 1994.

Roll-Hansen, Nils. "Experimental Method and Spontaneous Generation: The Controversy between Pasteur and Pouchet, 1859–1864." *Journal of the History of Medicine,* July 1979, vol. 34, pp. 273–292.

Taylor, P. A. *Current Fallacies about Vaccination: A Letter to Dr. W. B. Carpenter.* London: Allen, 1881.

Vallery-Radot, René. *The Life of Pasteur.* Garden City, N.Y.: Doubleday, Page, 1926.

Vallery-Radot, Pasteur. *Louis Pasteur: A Great Life in Brief.* New York: Alfred A. Knopf, 1958.

6 Golgi versus Ramón y Cajal

Addison, William H. F. "Ramón y Cajal—an Appreciation." *Scientific Monthly,* 1930, vol. 31, pp. 178–183.

Bergey, Gregory K. Review of *Histology of the Nervous System of Man and Vertebrates,* by Santiago Ramón y Cajal. *New England Journal of Medicine,* October 19, 1995, p. 1088.

Cannon, Dorothy F. *Explorer of the Human Brain: The Life of Santiago Ramón y Cajal (1852–1934).* New York: Henry Schuman, 1949.

Chen, Victor W. "Camillo Golgi, 1906." In Magill, 1991, vol. 1, pp. 79–86.

Clifford, Eth. *The Wild One.* Boston: Houghton Mifflin, 1974. (Fictionalized account of Ramón y Cajal's young life.)

E. A. S.-S. "Camillo Golgi." (Obituary notice.) *Proceedings of the Royal Society of Edinburgh,* 1925–1926, vol. 46, pp. 360–361.

Grant, Gunnar. "How Golgi Shared the 1906 Nobel Prize in Physiology or Medicine with Cajal." Online: http://www.nobel.se/hos/grant. (Based on a 1998 lecture at the Accademia di Medicina di Torino, Italy.)

Greenfield, Susan A. *The Human Brain: A Guided Tour.* New York: Basic Books, 1997. (For the nonspecialist.)

Kühnel, Wolfgang. *Pocket Atlas of Cytology, Histology and Microscopic Anatomy.* Rev. and enl. ed. 1950. Reprint, New York: Thieme Medical Publishers, 1992.

Loewy, Arthur D. "Ramón y Cajal and Methods of Neuroanatomical Research." *Perspectives in Biology and Medicine,* Autumn 1971, vol. 15, pp. 7–31.

Mazzarello, Paolo. *The Hidden Structure: A Scientific Biography of Camillo Golgi.* New York: Oxford University Press, 1999.

Minkoff, Eli C. "Santiago Ramón y Cajal, 1906." In Magill, 1991, vol. 1, pp. 89–95.

Ramón y Cajal, Santiago. *Recollections of My Life,* 2 vols. Trans. E. Horne Craigie. Philadelphia: American Philosophical Society, 1937. (This translation is of the third Spanish edition [1923] of *Recuerdos De Mi Vida,* 1901–1917.)

——. *Histology of the Nervous System of Man and Vertebrates.* Trans. Neely Swanson and Larry Swanson. 2 vols. History of Neuroscience, no. 7. New York: Oxford University Press, 1995.

——. *Advice for a Young Investigator.* Cambridge, Mass.: MIT Press, 1999.

Restak, Richard. *Brainscapes.* New York: Hyperion, 1995. (Guided tour of the brain, for the nonspecialist.)

Ross, Michael H., et al. *Histology: A Text and Atlas.* 2nd ed. Baltimore: Williams and Wilkins, 1989.

Rothman, James E. "The Compartmental Organization of the Golgi Apparatus." *Scientific American,* September 1985, pp. 74–88. (Background information on the Golgi apparatus.)

Rothman, James E., and Lelio Orci. "Budding Vesicles in Living Cells." *Scientific American,* March 1996, pp. 70–75. (Background information on the Golgi apparatus.)

Santini, M., ed. *Golgi Centennial Symposium: Proceedings,* New York: Raven Press, 1975.

Taylor, Douglass W. "Ramón y Cajal, Santiago." In Gillispie, 1975, vol. 7, pp. 273–276.

Zanobio, Bruno, "Golgi, Camillo." In Gillispie, 1972, vol., 5, pp. 459–461.

7. Freud versus Moll, Breuer, Jung, and Many Others

Amacher, Peter. "Freud, Sigmund." In Gillispie, 1972, vol. 5, pp. 171–181.

Bakan, David. *Sigmund Freud and the Jewish Mystical Tradition.* 1958. Reprint, New York: Schocken Books, 1965.

Bettelheim, Bruno. *Freud's Vienna and Other Essays.* New York: Alfred A. Knopf, 1990. (See especially "Freud's Vienna," pp. 3–17.)

Binger, Carl. *The Two Faces of Medicine: The Human Aspects and Relationships between Medicine and Psychiatry.* New York: W. W. Norton, 1967. (See especially chapter 12, "Freud and Medicine.")

Bloom, Harold, ed. *Sigmund Freud.* New York: Chelsea House, 1985.

Bollag, Burton. "Searching in Vienna for the Roots of Psychology." *Chronicle of Higher Education,* July 18, 1997, vol. 43, no. 5, p. B2.

Bower, Bruce. "Dr. Freud Goes to Washington." *Science News,* November 28, 1998, pp. 347–349.

———. "The Mental Butler Did It." *Science News,* October 30, 1999, pp. 280–282.

Bragg, Melvyn. *On Giants' Shoulders: Great Scientists and Their Discoveries from Archimedes to DNA.* New York: John Wiley and Sons, 1998.

Brill, A. A. *The Basic Writings of Sigmund Freud,* 1938. Reprint, New York: Modern Library, 1966.

Brome, Vincent. *Freud and His Early Circle.* New York: William Morrow, 1968.

Calne, Donald B. *Within Reason: Rationality and Human Behavior.* New York: Pantheon, 1999.

Cioffi, Frank. "Freud and the Idea of a Pseudo-Science." In Robert Borger, ed., *Explanation in the Behavioural Sciences.* Cambridge, Eng.: Cambridge University Press, 1970, pp. 471–499.

———. "Was Freud a Liar?" In Crews, 1998, pp. 34–42. (From a BBC radio interview, November 3, 1973.)

Clark, Ronald W. *Freud: The Man and the Cause.* New York: Random House, 1980.

Crews, Frederick C. *The Memory Wars: Freud's Legacy in Dispute.* New York: *New York Review of Books,* 1995.

———, ed. *Unauthorized Freud: Doubters Confront a Legend.* New York: Viking, 1998. (Blames Freud for the "recovered memory" fracas.)

Dimen, Muriel. "Strange Hearts: On the Paradoxical Liaison between Psychoanalysis and Feminism." In Roth, 1998, pp. 207–220.

Edmundson, Mark. "Save Sigmund Freud." *New York Times Magazine,* July 13, 1997, p. 34.

Fancher, Raymond E. *Pioneers of Psychology.* New York: W. W. Norton, 1979.

Freud, Sigmund. *An Autobiographical Study,* 1925, 1935. Reprint, New York: W. W. Norton, 1952.

Frey-Rohn, Lilians. *From Freud to Jung: A Comparative Study of the Psychology of the Unconscious.* 1974. Reprint, Boston: Shambala Publications, Inc., 1990. (A Jungian's point of view.)

Gay, Peter. "Sigmund Freud, Psychoanalyst." *Time,* March 29, 1999, pp. 66–69.

———, ed. *The Freud Reader.* New York: W. W. Norton, 1989.

Gilman, Sander L. *The Case of Sigmund Freud: Medicine and Identity at the Fin de Siècle.* 1993. Reprint, Baltimore: Johns Hopkins University Press, 1994.

Goode, Erica. "Return to the Couch: A Revival for Analysis." *New York Times,* January 12, 1999, pp. C1, C6.

——. "New Clues to Why We Dream." *New York Times,* November 2, 1999, pp. F1, F4.

Gruengard, Ora. "Introverted, Extroverted, and Perverted Controversy: Jung against Freud." *Science in Context,* Summer 1998, vol. 11, no. 2, pp. 255–290.

Grünbaum, Adolph. "A Century of Psychoanalysis: Critical Retrospect and Prospect." In Roth, 1998, pp. 183–195.

Horgan, John. "Why Freud Isn't Dead." *Scientific American,* December 1996, pp. 106–111.

Interview (staff). "Couch Wars." Michael S. Roth, Andrew Sullivan, and Muriel Dimen discuss the Freud exhibit at Washington's Library of Congress. *Interview,* October 1998, vol. 28, no. 10, p. 104.

Jelliffe, Smith Ely. "1920s; the Wizard of Dreams." Review of Freud's *A General Introduction to Psychoanalysis,* August 8, 1920. Reprinted in *New York Times Book Review,* October 6, 1996, p. 25.

Jones, Ernest. *The Life and Work of Sigmund Freud.* 3 vols. New York: Basic Books, 1953–1957. (See especially vol. 2, *Years of Maturity, 1901–1919,* published in 1955.)

Kiell, Norman, ed. *Freud without Hindsight: Reviews of His Work, 1893–1939.* Madison, Conn.: International Universities Press, 1988.

Kurzweil, Edith. "Did Freud Commit Fraud?" *Society,* March–April 1994, vol. 31, no. 3, p. 34. Obtained online: http://web5.infotrac.galegroup.com.

Laing, R. D. *Wisdom, Madness and Folly.* New York: McGraw-Hill, 1985.

Lear, Jonathan. *Open Minded: Working Out the Logic of the Soul.* Cambridge, Mass.: Harvard University Press, 1998. (See especially chapter 2, "On Killing Freud [Again]," pp. 16–32. This chapter was also published in slightly different form as "The Shrink: Is In: A Counterblast in the War on Freud," *New Republic,* December 25, 1995, pp. 18–25.)

Leland, John. "The Trouble with Sigmund." *Newsweek,* December 18, 1995, vol. 126, no. 25, p. 62. (On the postponement of the Library of Congress Freud exhibit.)

Lomas, Peter. Review of Gilman, 1995. *Society,* January–February 1996, vol. 33, no. 2, pp. 95, 96.

Malcolm, Janet. *In the Freud Archives.* New York: Alfred A. Knopf, 1984.

Masson, Jeffrey Moussaieff, ed. and trans. *The Complete Letters of Sigmund Freud to Wilhelm Fliess, 1887–1904.* Cambridge, Mass.: Belknap Press, 1985.

McDonald, Marci. "Burying Freud and Praising Him." *U.S. News and World Report,* October 19, 1998, pp. 60, 61.

McGuire, William, ed. *The Freud/Jung Letters: The Correspondence between Sigmund Freud and C. G. Jung.* Princeton, N.J.: Princeton University Press, 1974.

Merkin, Daphne. "Freud Rising." *New Yorker,* November 9, 1998, pp. 50–55. (Essay review of the Library of Congress exhibition.)

Miller, Jonathan, ed. *Freud: The Man, His World, His Influence.* Boston: Little, Brown, 1972.

Mlot, Christine. "Probing the Biology of Emotion (Unmasking the Emotional Unconscious)." *Science,* May 15, 1998, vol. 280, no. 5366, pp. 1005–1007.

Molnar, Michael, ed. and trans. *The Diary of Sigmund Freud, 1929–1939: A Record of the Decade.* New York: Charles Scribner's Sons, 1992.

Muckenhoupt, Margaret. *Sigmund Freud: Explorer of the Unconscious.* New York: Oxford University Press, 1997.

Niehoff, Debra. *The Biology of Violence: How Understanding the Brain, Behavior and Environment Can Break the Vicious Cycle of Aggression.* New York: Free Press, 1999.

Pollack, Robert. *The Missing Moment.* Boston: Houghton Mifflin, 1999.

Roazen, Paul. *Freud and His Followers.* New York: Alfred A. Knopf, 1975.

Rosenfeld, Israel. *Freud's Megalomania.* New York: W. W. Norton, 2000. (Rosenfeld creates a fictional manuscript as it might have been written by Freud.)

Roth, Michael S., ed. *Freud, Conflict, and Culture: Essays on His Life, Work, and Legacy.* New York: Alfred A. Knopf, 1998.

Schultz, Duane P. *Intimate Friends, Dangerous Rivals: The Turbulent Relationship between Freud & Jung.* Los Angeles: J. P. Tarcher, 1990.

Shorter, Edward. *A History of Psychiatry.* New York: John Wiley and Sons, 1997.

Sulloway, Frank J. *Freud, Biologist of the Mind: Beyond the Psychoanalytic Legend.* New York: Basic Books, 1979.

Tallis, R. C. "Burying Freud." *Lancet,* March 9, 1996, vol. 347, no. 9002, p. 669.

Torrey, E. Fuller. *The Death of Psychiatry.* Radnor, Pa.: Chilton, 1974.

Trillin, Calvin. "So, Nu, Dr. Freud?" *New Yorker,* August 15, 1999, p. 80.

Zaretsky, Eli. "The Attack on Freud." *Tikkun,* May–June 1994, vol. 9, no. 3, p. 65.

8. Sabin versus Salk

Black, Kathryn. *In the Shadow of Polio.* Reading, Mass.: Addison-Wesley, 1996.

Blume, Stuart, and Ingrid Geesink. "A Brief History of Polio Vaccines." *Science,* June 2, 2000, vol. 288, pp. 1593–1594.

Bredeson, Carmen. *Jonas Salk: Discoverer of the Polio Vaccine.* Hillsdale, N.J.: Enslow Publishers, 1993.

Burns, John F. "Vaccine War Emboldens India as It Weakens Polio." *New York Times,* January 26, 1997, p. 3.

Carter, Richard. *Breakthrough: The Saga of Jonas Salk.* New York: Trident Press, 1966.

Cimons, Marlene. "Polio's Nearly Extinct, but Sabin-Salk Feud Lives On." Los Angeles Times News Service, in *Hackensack Record,* March 13, 1983, p. A11.

Curson, Marjorie. *Jonas Salk.* Englewood Cliffs, N.J.: Silver Burdett, 1990. (Juvenile.)

Gould, Tony. *A Summer Plague: Polio and Its Survivors.* New Haven, Conn.: Yale University Press, 1995.

Grady, Denise. "As Polio Fades, Dr. Salk's Vaccine Re-emerges." *New York Times,* December 14, 1999, pp. F1, F5.

Klein, Aaron E. *Trial by Fury: The Polio Vaccine Controversy.* New York: Charles Scribner's Sons, 1972.

Nuland, Sherwin. Essay review of Gould, 1995. *New Republic,* October 16, 1995, vol. 213, no. 16, pp. 47–52.

Oldstone, Michael B. A. *Viruses, Plagues, and History.* New York: Oxford University Press, 1998. (See especially pp. 90–115.)

Podolsky, M. Lawrence. *Cures out of Chaos.* Amsterdam: Harwood Academic Publishers, 1998. (See chapter 9, "Orphans and Their Relatives," for background information on viruses.)

Roberts, Leslie. "Can We Kiss the Polio Scourge Goodbye?" *U.S. News and World Report,* August 30, 1999, p. 58.

Rogers, Naomi. *Dirt and Disease: Polio before FDR.* New Brunswick, N.J.: Rutgers University Press, 1992.

Salk, Jonas. "Studies in Human Subjects on Active Immunization against Poliomyelitis. 1. A Preliminary Report." *Journal of the American Medical Association,* March 28, 1953, pp. 1081–1098.

Seavey, Nina Gilden, Jane S. Smith, and Paul Wagner. *A Paralyzing Fear: The Triumph over Polio in America.* New York: TV Books, 1998.

Sheed, Wilfrid. "Jonas Salk." *Time,* March 29, 1999, vol. 153, no. 12, pp. 168–170.

Sherrow, Victoria. *Jonas Salk.* New York: Facts on File, 1993. (Juvenile.)

Smith, Jane. *Patenting the Sun: Polio and the Salk Vaccine.* New York: Morrow, 1990.

Tomlinson, Michael. *Jonas Salk.* Vero Beach, Fla.: Rourke Publications, 1993. (Juvenile.)

Weisse, Allen B. *Medical Odysseys: The Different and Sometimes Unex-*

pected Pathways to Twentieth-Century Medical Discoveries. New Brunswick, N.J.: Rutgers University Press, 1991. (See "Polio: The Not-So-Twentieth-Century Disease," pp. 158–185.)

Zinsser, Hans. *Rats, Lice and History.* 1935. Reprint, New York: Black Dog and Leventhal, 1996.

9. Franklin versus Wilkins

Bate, Jon, and Hilary Gaskin. "Unsung Pioneer." *New Statesman,* July 8, 1983, vol. 106, no. 2729, p. 12.

Bernal, J. D. "Dr. Rosalind E. Franklin." *Nature,* July 19, 1958, p. 154.

Bernstein, Jeremy. *Experiencing Science. Profiles in Discovery.* New York: Basic Books, 1978.

Bragg, Melvyn. *On Giants' Shoulders: Great Scientists and Their Discoveries from Archimedes to DNA.* New York: John Wiley and Sons, 1998.

Chargaff, Erwin. "A Quick Climb Up Mount Olympus." *Science,* March 29, 1968, vol. 159, no. 3822, pp. 1448, 1449. (Review of *The Double Helix,* by James Watson.)

———. *Heraclitean Fire: Sketches from a Life before Nature.* New York: Rockefeller University Press, 1978.

Colomé, Jaime S. "James D. Watson, 1962." In Magill, 1991, vol. 2, pp. 849–860.

Crick, Francis. *What Mad Pursuit: A Personal View of Scientific Discovery.* New York: Basic Books, 1988.

Curtis, Robert H. *Medicine.* New York: Scribner's, 1993. (See especially the chapter on Watson and Crick, pp. 250–260.)

Franklin, R. E., and R. G. Gosling. "Molecular Configuration in Sodium Thymonucleate." *Nature,* April 25, 1953, vol. 171, no. 4356, pp. 740–741.

Friedman, Meyer, and Gerald W. Friedland. "Maurice Wilkins and DNA." In *Medicine's 10 Greatest Discoveries.* New Haven, Conn.: Yale University Press, 1998, pp. 192–227.

Garber, Ken. "High Stakes for Gene Therapy." *Technology Review,* March–April 2000, vol. 103, no. 2, pp. 58–60, 62, 64.

Henderson, Bennye S. "Wilkins, Maurice H. F., 1962." In Magill, 1991, vol. 2, pp. 861–868.

Hoagland, Mahlon. *Discovery: The Search for DNA's Secrets.* Boston: Houghton Mifflin, 1981.

Hollar, David Wason, Jr. "Francis Crick, 1962." In Magill, 1991, vol. 2, pp. 839–847.

Judson, Horace Freeland. *The Eighth Day of Creation: The Makers of the Revolution in Biology.* New York: Simon and Schuster, 1979.

————. "Annals of Science: The Legend of Rosalind Franklin." *Science Digest,* January 1986, pp. 56–59, 78–83.

Kass-Simon, G., and Farnes, Patricia, eds. *Women of Science: Righting the Record.* Bloomington: Indiana University Press, 1990. (See pp. 335–348 and 359–364.)

Lagerqvist, Ulf. *DNA Pioneers and Their Legacy.* New Haven, Conn.: Yale University Press, 1998.

Licking, Ellen F. "Double-Teaming the Double Helix." *U.S. News and World Report,* August 17, 1998, vol. 125, no. 7, pp. 72–74.

Lwoff, André. "Truth, Truth, What Is Truth (About How the Structure of DNA Was Discovered)?" *Scientific American,* July 1968, vol. 219, no. 1, pp. 133–137. (Essay review of *The Double Helix,* by James Watson.)

Olby, Robert. "Francis Crick, DNA, and the Central Dogma." *Daedalus,* Fall 1970, pp. 938–987. (Issued as vol. 99, no. 4 of the *Proceedings of the American Academy of Arts and Sciences.*)

————. "Franklin, Rosalind Elsie." In Gillispie, 1972, vol. 5, pp. 139–142.

————. *The Path to the Double Helix.* Seattle: University of Washington Press, 1974.

Pauling, Linus. "Molecular Basis of Biological Specificity." *Nature,* April 26, 1974, vol. 248, pp. 769–771.

Perutz, Max F. "DNA Helix." (Letter commenting on Chargaff's review of *The Double Helix,* by James Watson.) *Science,* June 27, 1969, vol. 164, no. 3887, pp. 1537–1538.

Portugal, Franklin H., and Jack S. Cohen. *A Century of DNA: A History of the Discovery of the Structure and Function of the Genetic Substance.* Cambridge, Mass.: MIT Press, 1977.

Sayre, Anne. *Rosalind Franklin and DNA: A Vivid View of What It Is Like to Be a Gifted Woman in an Especially Male Profession.* New York: W. W. Norton, 1975.

Stent, Gunther S. "DNA." *Daedalus,* Fall 1970, pp. 909–937. (Issued as vol. 99, no. 4 of the *Proceedings of the American Academy of Arts and Sciences.*)

Watson, J. D. *The Double Helix.* New York: Atheneum, 1968.

————. Letter commenting on Erwin Chargaff's review of *The Double Helix,* by James Watson. *Science,* June 27, 1969, vol. 164, no. 3887, p. 1539.

Watson, J. D., and F. H. C. Crick. "A Structure for Deoxyribose Nucleic Acid." *Nature,* April 25, 1953, vol. 171, no. 4356, pp. 737–738.

Wilkins, M. H. F. Letter commenting on Erwin Chargaff's review of *The Double Helix,* by James Watson. *Science,* June 27, 1969, vol. 164, no. 3887, p. 1539.

Wilkins, M. H. F., A. R. Stokes, and H. R. Wilson. "Molecular Structure of

Deoxyribose Nucleic Acids." *Nature,* April 25, 1953, vol. 171, no. 4356, p. 738–740.

10. Gallo versus Montagnier

Altman, Lawrence K. "New Book Challenges Theories of AIDS Origins." *New York Times,* November 30, 1999, pp. F1, F6.

Armelagos, George J. "The Viral Superhighway." *The Sciences,* January–February 1998, pp. 24–28. (Background on viruses.)

Clines, Francis X. "For Besieged Scientist, New Start in New Lab." *New York Times,* March 11, 1997, p. C1.

Cohen, Jon. "John Crewdson: Science Journalist as Investigator." *Science,* November 15, 1991, vol. 254, pp. 946–949.

——. "Report Card on Crewdson's Reporting." *Science,* November 15, 1991, vol. 254, pp. 948–949.

——. "Pasteur Wants More HIV Blood Test Royalties." *Science,* February 14, 1992, p. 792.

——. "HHS: Gallo Guilty of Misconduct." *Science,* January 8, 1993, vol. 259, pp. 168–170.

——. "U.S.-French Patent Dispute Heads for a Showdown." *Science,* July 1, 1994, vol. 265, pp. 23–25.

——. "The Duesberg Phenomenon." *Science,* December 9, 1994, vol. 266, pp. 1642–1644.

Cohen, Jon, and Eliot Marshall. "NIH-Pasteur: A Final Rapprochement." *Science,* July 15, 1994, vol. 265, p. 313.

Duesberg, Peter H. *Inventing the AIDS Virus.* Washington, D.C.: Regnery, 1996.

Economist (Staff). "The One True Virus." *Economist,* June 8, 1991, pp. 83–84.

Epstein, Steven. *Impure Science: AIDS, Activism, and the Politics of Science.* Berkeley: University of California Press, 1996.

Gallo, Robert. *Virus Hunting: AIDS, Cancer and the Human Retrovirus, a Story of Scientific Discovery.* New York: Basic Books, 1991.

Gallo, Robert, in discussion with Paul Hoffman. *Viruses: The Greatest Threat to the Survival of Our Species.* Videotape. San Ramon, Calif.: Pangea Digital Pictures, 1995.

Gorman, Christine. "Victory at Last for a Besieged Virus Hunter." *Time,* November 22, 1993, vol. 142, no. 22, p. 61.

Greenberg, Daniel S. "Resounding Echoes of Gallo Case." *Lancet,* March 11, 1995, vol. 345, p. 639.

Hamilton, David P. "OSI Investigator 'Reined In.' " *Science,* July 26, 1991, vol. 253, p. 372.

———. "What Next in the Gallo Case?" *Science,* November 15, 1991, vol. 254, pp. 944–945.

Heginbotham, Stanley J. "The Power of HIV-Positive Thinking." *Sciences,* May–June 1997, pp. 38–42. (Essay review of Epstein, 1996.)

Hooper, Edward. *The River: A Journey to the Source of HIV and AIDS.* Boston: Little, Brown, 1999.

Hooper, Judith. "A New Germ Theory." *Atlantic Monthly,* February 1999, pp. 41–49.

Horowitz, Craig. "Winning the Cancer War." *New York,* February 7, 2000, pp. 26–31.

Horton, Richard. "Truth and Heresy About AIDS." *New York Review of Books,* May 23, 1996, pp. 14, 16–20. (Essay review of three books written or edited by Peter Duesberg.)

Jasanoff, Sheila. "An Epidemic Challenges American Biomedicine." *American Scientist,* September–October 1997, vol. 85, pp. 476–477. (Review of Epstein, 1996.)

Kaiser, Jocelyn. "The 'Gallo Case': Popovic Strikes Back." *Science,* February 14, 1997, vol. 275, pp. 920, 921.

Klatzmann, David, et al. "Immune Status of AIDS Patients in France: Relationship with Lymphadenopathy Associated Virus Tropism." *Annals of the New York Academy of Sciences,* December 29, 1984, pp. 228–237. (Authors include Jean-Claude Chermann and Luc Montagnier.)

Leibowitch, Jacques. *A Strange Virus of Unknown Origin.* Trans. Richard Howard. Introd. Robert Gallo. New York: Ballantine, 1985.

Markham, Phillip D., et al. "Correlation between Exposure to Human T-Cell Leukemia-Lymphoma Virus-III and the Development of AIDS." *Annals of the New York Academy of Sciences,* December 29, 1984, pp. 106–109. (Authors include Mika Popovic and Robert Gallo.)

Montagnier, Luc. *Virus: The Co-Discoverer of HIV Tracks Its Rampage and Charts the Future.* New York: W. W. Norton, 2000. (Translated by Stephen Sartavelli from French edition, 1994.)

Oldstone, Michael B. A. *Viruses, Plagues, and History.* New York: Oxford University Press, 1998. (See pp. 140–157.)

Palca, Joseph. "Hints Emerge from the Gallo Probe." *Science,* August 16, 1991, vol. 253, pp. 728–731.

Radetsky, Peter. "Immune to a Plague." *Discover,* June 1997, vol. 18, no. 6, pp. 60 ff. (Obtained online.)

Random House Webster's Dictionary of Scientists. New York: Random House, 1997.

Rubinstein, Ellis. "The Gallo Factor: Questions Remain." *Science,* August 16, 1991, vol. 253, p. 732.

Schoub, Barry D. *AIDS and HIV in Perspective: A Guide to Understanding the Virus and Its Consequences.* Cambridge, Eng.: Cambridge University Press, 1994.

Shilts, Randy. *And the Band Played On: Politics, People, and the AIDS Epidemic.* 1987. Reprint, New York: Penguin Books, 1988.

Stone, Richard. "Dingell Pursues AIDS Patent 'Cover-Up.' " *Science,* July 30, 1993, vol. 261, no. 5121, p. 539.

INDEX